Ammonites

Neale Monks and Philip Palmer

Smithsonian Institution Press, Washington, D.C.
in association with The Natural History Museum, London

For VJM

Picture credits:
Mike Eaton/© NHM: pp. 16, 34-5, 37, 39, 44, 58, 63–4, 100
© Victoria Edwards/NHM: p.49
© John Sibbick/NHM: p.67 top

All other images are copyright of The Natural History Museum, London and taken by the Museum's Photo Unit.

For copies of these and other images please contact The Picture Library, The Natural History Museum, Cromwell Road, London SW7 5BD
View the Picture Library website at www.nhm.ac.uk/piclib

© The Natural History Museum, London, 2002

All rights reserved. No part of this publication may be reproduced or transmitted in any form or by any means without prior written permission from the British publisher.

Published in the United States of America
by the Smithsonian Institution Press
in association with The Natural History Museum, London
Cromwell Road
London SW7 5BD
United Kingdom

Library of Congress Cataloging-in-Publication Data
Monks, Neale.
    Ammonites / Neale Monks and Philip Palmer.
        p.   cm.
    Includes bibliographical references (p.).
    ISBN 1-58834-024-4 (alk. paper)—ISBN 1-58834-047-3 (pbk. : alk. paper)
        1. Ammonoidea. I. Palmer, Philip. II. Title.
    QE807.A5 M68   2002
    564'.53—dc21   2001049552

Manufactured in Singapore, not at government expense

09 08 07 06 05 04 03 02   5 4 3 2 1

Edited by Rebecca Harman
Designed by Mercer Design
Reproduction and printing by Craft Print, Singapore

Front cover main image: *Parkinsonia dorsetensis*, Middle Jurassic ammonite

Contents

Preface   5

Acknowledgements   7

Chapter 1: An introduction to ammonites   9
   The Cephalopoda
   A short history of the Cephalopoda
   A clue to the past: *Nautilus*
   The anatomy of *Nautilus*

Chapter 2: Ammonite fossils   33
   Ammonite shell morphology
   The suture line
   The aptychus
   Ammonite soft body anatomy

Chapter 3: Ammonite form and function   53
   Buoyancy
   Water pressure
   Orientation
   Jet propulsion, half a billion years before Frank Whittle
   Streamlining
   Defence

Chapter 4: Aspects of ammonite biology   89
  Sexual dimorphism
  Ammonite reproduction
  Ammonite old age, pathology and predators

Chapter 5: Ammonite taxonomy and classification   107
  Suborder Anarcestina
  Suborder Clymeniina
  Suborder Goniatitina – the goniatites
  Suborder Prolecanitina
  Suborder Ceratitina – the ceratites
  Suborder Phylloceratina
  Suborder Lytoceratina
  Suborder Ancyloceratina – the heteromorphs
  Suborder Ammonitina – the true ammonites

Chapter 6: The extinction of the ammonites   135
  Periodic extinctions throughout the Palaeozoic and Mesozoic
  Late Cretaceous decline and extinction
  The ammonites' successors – the Coleoidea

Collecting ammonites and ammonite collections   143
Further information   145
Glossary   148
Index   154

Preface

Most of what is written about ammonites is found in scientific journals, and an account of the natural history of ammonites for the general reader has long been overdue. The last such book, that went some way to bringing ammonites to the general reader was *The ammonites: Their life and their world,* was written by Ulrich Lehman and translated into English in 1981. Lehman drew from the latest scientific papers of the time, many of which he wrote himself, creating an impressive array of facts and theories which produced a detailed but very readable book. However, since its publication, palaeontologists have not just accumulated more information about ammonites but have also made some fundamental changes in the way ammonites are envisaged. The most significant has been the switch away from the nautilus as the prototype of the ammonite animal towards the more active cephalopods like the octopus. Any new book on ammonites must take into account these new observations and ideas, and in this book we describe some of the most important ones.

This book is not about the geology of ammonite fossils. Lists of stratigraphical and geographical occurrences are absent, and not much is said about the use of ammonites in stratigraphy, important as that is. Also absent are involved descriptions of ammonite taxonomy, which would easily fill a book this size several times over. Instead, we hope that by placing the accent on the biology of ammonites, we have provided enough to get the student, the enthusiastic amateur, or the armchair naturalist started in the tricky field of ammonite palaeoecology. Not all the theories described here are held by every ammonite expert, but we make no apology for that. Ammonites may be abundant fossils, but with regard to some important questions they offer very little in the way of solid evidence. Indeed, when it comes to ammonites, palaeontologists have been in agreement on remarkably little! By laying before the reader some of these questions, and the facts available to answer them, we hope to show that despite being familiar fossils, ammonites are not well understood at all.

Note that in this book 'ammonite' is taken to be any member of the order Ammonoidea, though strictly speaking 'ammonite' is used for the single suborder Ammonitina within the Ammonoidea (see p. 128).

Acknowledgements

The authors would like to thank all of their many friends and colleagues at The Natural History Museum who helped and encouraged the authors along the way. In particular, the assistance of the Curator of Fossil Cephalopods, Steve Baker, is noted with thanks; without his help in examining the specimens in his care, this book would not have been possible. The authors also wish to thank Robert Ralph for reading through the manuscript and providing many useful comments and suggestions. It was as an undergraduate at the University of Aberdeen that one of us (NM) was inspired by Bob to take writing seriously! The authors would also like to thank two anonymous reviewers from the USA. NM would also like to acknowledge the frequent collaborations and conversations with other palaeontologists that went a long way towards defining his understanding of ammonites (though of course the opinions of the authors cited here are their own!). Foremost on this list include Ottilia Szives, W. J. Kennedy, Andy Gale, Andrew Smith, Noel Morris and Jeremy Young, but there are many, many others.

Chapter One

AN INTRODUCTION TO AMMONITES

AMMONITES intrigue people who know nothing about fossils, just as much as they do amateur and professional palaeontologists. Amateur fossil collectors quickly build up collections of dozens of species of ammonites, while the mathematically-minded delight in the precision of their spiral shells. Gift shops around the world frequently sell ammonites as attractive ornaments for the home, from small ones costing very little right up to giant ones for the seriously wealthy connoisseur. The compact, spirally coiled shells of ammonites have inspired artisans and architects. For example, the house of the Victorian naturalist Gideon Mantell in Lewes, Sussex, features classical Ionic columns, and the whorls at the tops of the columns are not plain spirals but stone replicas of the mid-Jurassic ammonite *Teloceras*.

The ammonites have contributed to mythology and place names to a surprisingly high degree. In England, ammonites were called 'snakestones', and in some places snake-like heads were even carved onto them. One particular town with a legend explaining the snakestones is Whitby, on the Yorkshire coast. The story goes that the town was plagued by snakes until the Abbess, St. Hilda, turned them all into stone. The cliffs around the town expose Lower Jurassic shales that are particularly rich in fossil ammonites,

left: Lower Jurassic snakestone from Whitby, Yorkshire, UK. A snake's head has been carved on to the ammonite *Dactylioceras commune*.

above: The Upper Jurassic ammonite *Virgatosphinctes* sp. from the Himalayas.

including one that palaeontologists call *Hildoceras*. In India, certain parts of the Himalayas are made up of Jurassic rocks, and periodically rivers wash out fossil ammonites. The local people hold these fossils in great esteem. They see them as symbols of the god Vishnu, the supreme god of the Hindus. On wedding days, such an ammonite is a traditional gift from the bride's family to the bridegroom, and it becomes a central part of the household's rituals and prayers.

Delightful as the artistic, mythological and religious associations of ammonites are, the reality is much more down to earth. Ammonites are cephalopods, relatives of the nautiluses, octopuses, squids and cuttlefish. Since they are extinct, it is only through study of their fossils that palaeontologists have been able to piece together something of their anatomy, their habitat and their ecology. They are a kind of mollusc, characterised by having an external shell divided into chambers by complicated partitions called septa. Molluscs share a number of distinct features. All are soft bodied, and most have an external, calcareous shell secreted by a special layer of tissue called the mantle. Most are aquatic and even the terrestrial snails and slugs prefer damp habitats and cannot tolerate dry conditions for very long. Although they eat a wide variety of foods, most are equipped with a rasping tongue called the radula which is covered in rows of teeth. Molluscs can be divided into a number of classes all sharing a common ancestor believed to have lived early in the Cambrian period about 550 million years ago. By far the largest group is the class Gastropoda, which includes the snails and slugs, of which there are approximately 50,000 species. The next largest class is the Bivalvia of which there are over 10,000 species alive today. These are the clams, oysters, scallops and mussels. Most live in the sea, although there are also some living in freshwater. The third major class is the Cephalopoda, to which the ammonites belong.

THE CEPHOLOPODA

The name Cephalopoda literally means 'head-footed', a reference to the ring of arms that all cephalopods have surrounding the mouth. Although sharing the same basic design as the other molluscs, cephalopods have become adapted to an active, predatory lifestyle. They are able to swim, and have muscular arms and a sharp beak for catching and dismembering their prey. Thousands of fossil species of cephalopod are known, of which the nautiluses, ammonites and belemnites are the best known groups. Only a

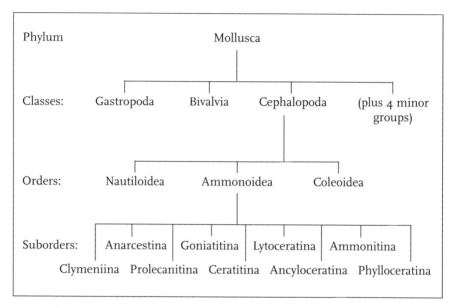

above: The three orders of cephalopods: the Nautiloidea, Ammonoidea and Coleoidea.

few hundred species of cephalopod are alive today, mostly squids, octopuses and nautiluses. There are no living ammonites or belemnites; they became extinct at the end of the Cretaceous.

Scientists have divided the cephalopods into three subclasses or orders, the Nautiloidea, Ammonoidea and Coleoidea. The Nautiloidea contains the living and fossil nautiluses. These have external shells, usually straight in the most primitive species but tightly coiled in the later ones, including the living species. Their shells are chambered, divided by gently curved walls called septa. On the radula, each row of teeth consists of one large central tooth and six small teeth on either side, making 13 teeth per row in total. The Ammonoidea includes the ammonites and their relatives such as the goniatites and ceratites. Superficially they resemble nautiluses in having a coiled, chambered shell, but ammonites have complex septa, with numerous folds and divisions at the edges where the septum meets the outer wall of the shell. The radula is also different, instead of 13 teeth there are only nine teeth per row, one central tooth and four on either side. The structure of ammonite shells, and their similarities and differences with nautiluses are discussed in more detail in Chapter 2.

The third group, the Coleoidea, includes the belemnites, the cuttlefish, the squids and the octopuses. Apart from the belemnites, this group has reached its greatest diversity in the modern seas and oceans, and examples can be found today in rock pools, coral reefs and throughout the depths of the ocean and abyssal plain. The Coleoidea have internal shells, sometimes chambered, as with cuttlefish, and other times not, as with squid. The octopuses have no shell except the argonauts (see p. 21). The radula of coleoids is similar to that of ammonites, with nine teeth, which suggests that coleoids and ammonoids are more closely related to one another than to the nautiluses. Unlike either the nautiluses or (as far as we can tell) the ammonites, coleoids can produce ink, which they use as a 'smoke-screen' to cover their escape when threatened. This has given rise to the common name for this group, the inkfish. Coleoids are also particularly intelligent compared with other invertebrate animals. They have sophisticated hunting, social and mating behaviours, and many species show an aptitude for learning. Octopuses, for example, have learned how to open bottles or press buttons to obtain morsels of food.

All nautiluses and ammonites, and the more primitive coleoids such as the belemnites and cuttlefish, share the same basic shell design. The shell is made from a mineral called aragonite, sometimes called mother-of-pearl, and, as noted above, is divided up into a series of chambers by the septa. Connecting the chambers is a strand of tissue called the siphuncle which passes through perforations in each of the chamber walls. Ultimately it joins the main body of the animal. The siphuncle is well supplied with blood, and is the transport system which pulls water out of the chambers while filling them up with gas. Emptying the chambers of water reduces the overall density of the animal, enabling it to attain neutral buoyancy, so it can float easily and swim with the minimum of effort. In terms of function it serves the same purpose as the swim bladder of bony fishes. The siphuncle can compensate for changes in pressure if the animal swims into shallower or deeper water, and is also able to pump out any water that seeps back into the shell. Ammonite shells are so similar to the shells of nautiluses, that it is certain they served the same purpose. The only differences are the more complex septa in the ammonites and the fact that the siphuncle does not run through the centre of the chambers, as in nautiluses, but usually near the bottom of chambers. The way in which ammonite shells worked, and why they had the shapes they did, are discussed in more detail in Chapter 3.

Though their shells are common fossils, there is very little evidence for the other aspects of ammonite anatomy. The radula has been found in quite a few species, and the siphuncle is often preserved. Strange structures called aptychi are commonly found associated with ammonites, though precisely what they were is open to debate. Some experts consider them to be trapdoors which sealed off the shell when the ammonite was threatened, while others believe them to be part of the jaws. Ammonite hatchlings, called ammonitellas, and even eggs are commonly found, and are about the same size as the eggs and hatchlings of modern squid. Occasional fossils have been found to show traces of the eyes, gills and tentacles, and the diet of a few species is known from what appear to be traces of the gut or stomach with fragments of food preserved. Therefore something of the ecology of ammonites can be infered from these few fossils, as reviewed in Chapter 4, but not nearly as much as palaeontologists would like. Studies of living cephalopods, particularly the coleoids, remain the best way to understand what ammonites might have been like.

If our knowledge of ammonite biology is limited, the same cannot be said about their diversity. Ammonite fossils are so common and widespread that they have become one of the most useful invertebrate groups for dating late Palaeozoic and Mesozoic rocks. Most importantly, a single species or subspecies often only lasted for 2 million years or less. If such a species can be found in rocks in different places, then geologists can say with some confidence that both formations were deposited at the same time. Many species of ammonites had a global distribution, making them ideal time 'markers'. Correlating different geological formations in time using fossils is a major part of a process called biostratigraphy. Biostratigraphy is very important in producing geological surveys and maps, and essential to mineral and oil prospecting. This is the main reason that ammonites have been so intensively studied. Because so many species of ammonites have been found and named, an elaborate system of classification has developed to order them. To cover the classification of ammonites would easily fill a book of this size, so instead a brief survey of the main groups and how they are related is given in Chapter 5.

A SHORT HISTORY OF THE CEPHALOPODA

THE NAUTILOIDEA

The first molluscs appeared during the early Cambrian, and among them were the earliest cephalopods, primitive nautiloids. Typical of these early nautiloids was *Plectronoceras*, which had a slightly curved, conical shell a bit like a tall limpet. Although such a poorly streamlined shell would not have been well adapted for swimming, the shell of *Plectronoceras* had both chambers and a siphuncle, and so presumably was able to regulate its buoyancy. Perhaps it drifted in the plankton like a jellyfish. Quite how and why these early cephalopods acquired chambered shells and the ability to control their buoyancy is not known, but by the Ordovician, nautiloids had become streamlined and capable swimmers. Long, straight shells, known as orthoconic shells, reduced drag and so allowed them to reach greater speeds with less effort. From the Ordovician onwards, all nautiloids had a notch in the lower edge of the opening of the shell called the hyponomic sinus, which indicates that they had evolved the system of jet propulsion still employed by all living cephalopods. The details of cephalopod jet propulsion are described in Chapter 3, but essentially they use a two-stroke system. Water is sucked into the body cavity inside the shell from outside, and then squeezed out through an adjustable tube, which in nautiluses is called the hyponome. The hyponomic sinus is the inflection in the lower edge of the shell through which the hyponome pokes, and so its presence in fossil nautiloids implies a hyponome, and therefore jet propulsion.

One peculiarity of jet propulsion as employed by cephalopods is that normally the jet emerges from the front of the shell, thus it sends the animal in the other direction, i.e. backwards. Being flexible, the hyponome can be used to direct the jet, and so provide some steering ability; however, forcing the jet through a tightly curved hyponome reduces its efficiency.

right: *Nautilus pompilius* showing the hyponomic sinus where the hyponome protrudes and the optic sinus.

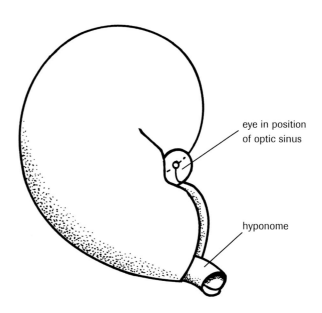

eye in position of optic sinus

hyponome

Living nautiluses move forwards very slowly, and their escape response is always to swim backwards, when the jet is at its most effective. Squids and cuttlefish rarely use the jet for swimming forwards, preferring to use their fins for that purpose, and like the nautiluses, reserve the jet for hasty retreats from danger. The octopus rarely swims, but when it does, it swims backwards.

As neutral buoyancy effectively made them weightless, these orthocone nautiloids were among the first really big animals. Most are between 30 cm and 2 m (12 in and 6.5 ft) long, far larger than any of the other invertebrates around at the time, and a few were substantially larger. The biggest were at least 5 m (16 ft) long, and as the top predators of their day must have been comparable to modern great white sharks or killer whales. One peculiarity of many orthocone nautiloids was that the chambers nearest the apex of the shell were filled, not with gas, but with a mineral called calcite, and as the nautiloid grew, successively more chambers were so filled. These solid chambers acted as counterweights for the main body of the animal.

above: The straight and chambered shell of the Cambrian nautiloid *Endoceras* sp.

During the Palaeozoic and Mesozoic, other groups of nautiloids appeared which had coiled rather than straight shells. Such shells are very compact, and bring the centres of mass and buoyancy together. The centre of mass (also known as the centre of gravity) is the point through which the weight of the animal acts, and the centre of buoyancy is the point through which its flotation acts. If they are close together, the animal is unstable, which makes them much more manoeuvrable. A small force will tip the shell up or down, or side to side. In contrast, the shells of the orthocone nautiloids have widely separated centres of mass and buoyancy. They would have been very stable and resistant to changes in direction, so while the orthocone shell shape was ideal for rapid backwards swimming, it was also very difficult to steer.

These coiled nautiloids steadily diversified throughout the Palaeozoic and Mesozoic, while the orthocone nautiloids became less common. Possibly the orthocone nautiloids could not compete with the rapidly evolving bony fish and sharks, which combined both speed and agility, and the last species of orthocone nautiloid became extinct in the Triassic. During the Mesozoic nautiluses were widespread but always less diverse than the ammonites, although the mass extinctions at the end of the Cretaceous which wiped out the ammonites, hardly affected the coiled nautiluses.

The Ammonoidea

The ancestors of the ammonites were a small and rather poorly known order of orthocone nautiloids known as the Bactritina. Even at their most diverse in the Devonian, they were not common, and many palaeontologists have never even heard of them, let alone seen one. It is perhaps surprising that this very obscure group of fossils should have given rise to one of the best known, the ammonites!

What distinguishes the bactritids from nautiloids is the globular shape, calcareous shell and small size of the first chambers, the so-called

above: The straight nautiloid *Bactrites carinatus* from the Devonian of Germany.

'embryonic shell'. Nautiloids, both fossil and recent, all have relatively large embryonic shells, and living species lay a few large eggs. In contrast, bactritids produced large numbers of small eggs and so did their descendants, the ammonites. Like bactritids and coleoids, ammonites had small embryonic shells, thus indicating a close relationship between the coleoids and ammonites.

The ammonites proper appeared early during the mid-Devonian. Unlike the bactritids they did not have orthoconic shells, but coiled shells like the coiled nautiloids, a striking example of convergent evolution. They probably evolved coiled shells for exactly the same reason as the nautiloids, improved manoeuvrability. By the end of the Devonian, the bactritids had all but become extinct, while the ammonites prospered. For the rest of the Palaeozoic and the entire Mesozoic, a period of well over 290 million years, the ammonites were the pre-eminent cephalopod group, and among the most diverse and widespread of all marine invertebrates. Why were they so successful? This is very difficult to say, but perhaps it was their ability to reproduce quickly that allowed them to evolve rapidly and so adapt to changing conditions. Until their gradual decline to extinction at the end of the Cretaceous, at least some ammonites survived all the previous major mass extinctions. These survivors diversified and gave rise to whole new families of ammonites; indeed, ammonites are among the fastest evolving animal groups we know about from the fossil record.

THE COLEOIDEA

A second group of cephalopods were derived from the bactritids; the coleoids. Although they share the bactritid and ammonite characteristic of having a small embryonic shell and a radula with nine teeth per row, they are different in having internal, rather than external, shells. In the more primitive groups the shell is similar to that of an orthocone bactritid, except in being internal. Functionally, these shells worked in the same sort of way, with gas-filled chambers connected by the siphuncle. Other groups of coleoids still use the shell to provide buoyancy, but in a modified form, such as the 'cuttlebone' possessed by cuttlefish. Many living species have reduced shells without chambers that do not provided buoyancy (squids) or else have

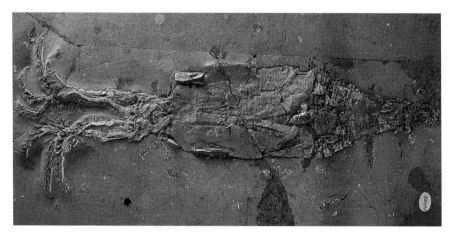

above: The Upper Jurassic *Belemnoteuthis antiquus*, a squid-like coleoid with hooks on its eight arms.

no shell at all (octopuses). Quite why these cephalopods have dispensed with the buoyant shell is not clear, perhaps it is related to living in deep water, where a hollow, gas-filled shell can be difficult to build and maintain.

The Coleoidea appeared in the Devonian, but did not become common until much later. The Jurassic was the heyday of first big group of coleoids, the belemnites, as well as the time when the first squids evolved. Also common in the Jurassic were vampyromorphs, squid-like animals of which only a single species survives today, confined to the deep sea. Octopuses, lacking any kind of shell, are rare fossils, but some are known from the Cretaceous, and cuttlefish appeared in the Cainozoic. Coleoids, such as the giant squid are the dominant cephalopod group and important members of many marine ecosystems.

As the successors of the ammonites, the evolution of the coleoids falls outside the scope of this book, but there is one kind which may have a direct connection with the evolution of the ammonites. This is the paper 'nautilus', or argonaut *(Argonauta* species), in fact a kind of octopus. Unlike most other octopuses, argonauts live close to the surface of warm seas rather than on the sea floor. The female produces a paper-thin egg case resembling a shell into which she lays her eggs. Compared with the females, the males are tiny, and while the females have been known since ancient times, the males were only described in the nineteenth century. The most curious thing about the argonaut is its egg case, particularly the fact that it closely resembles an ammonite shell, such as the Triassic ammonite *Trachyceras*. This is unlikely to be a result of convergent evolution, that is, a similar solution to a common problem, because the ammonite's shell and the argonaut's egg-case do completely different things. Could it be mere coincidence?

Female octopuses brood their eggs in caves. Living hundreds of metres above the sea floor, the argonaut cannot do this. It has been suggested that when

above: From the outside the egg case of *Argonauta hians* closely resembles the ammonite *Trachyceras* from the Triassic.

argonauts evolved in the Mesozoic they instead used the floating ammonite shells, left behind after the ammonite died, in which to brood their eggs. Some of these early argonauts even evolved ways of patching up damaged ammonite shells by secreting calcite from their arms, and eventually got so good at it they could make a facsimile of an ammonite shell from scratch! A second theory is more profound since it ties the argonauts much more closely to the ammonites than mere squatters of empty shells. Lewy (1996) suggested that argonauts are in fact 'nude ammonites', the culmination of a line of ammonites that switched from using the shell as a buoyancy device to an egg case. Lewy points to the strong sexual dimorphism of ammonites; like argonauts, ammonite species occur in two sizes, a big 'macroconch' and a small 'microconch'. He believes that the macroconch was the female, and that she had a larger shell so that she could brood the eggs safely inside. Yet again we have a nice story, with only circumstantial evidence to support it, but in its favour is one particular fact that is otherwise difficult to explain. Like other coleoid cephalopods, octopuses have nine teeth per row on the radula. What is striking is that in many species, particularly the more primitive deep-sea species, these radula teeth are very like those found in fossil ammonites, and not at all like those of other coleoids such as squid and cuttlefish. Is this a clue to a closer affinity than most palaeontologists care to imagine?

A CLUE TO THE PAST, *NAUTILUS*

Although coleoids share some features with ammonites, as far as the shell is concerned the two groups are very different. In contrast, the shells of nautiluses and ammonites are broadly similar, and fortunately for palaeontologists, a number of species of nautilus are still to be found in the tropical Indian and Pacific oceans today. They are usually all placed in a single genus, *Nautilus*, although recent studies have suggested there may be two distinct stocks, and so a second genus, *Allonautilus*, has been described. All the living species of nautilus are found in deep water, generally between 200 and 400 m (650 and 1310 ft) deep, particularly favouring the steep 'cliffs' that lie between islands and the deep sea floor. Nautiluses are opportunistic feeders, scavenging for large invertebrates like crustaceans.

Nautilus shells are strikingly beautiful, with a distinctive tiger stripe design on the outside, and an iridescent sheen on the inside, from which they get their common name of 'pearly nautilus'. Since the shell is buoyant, after the death and decay of the animal, the shell can float for long periods of time and wash up on a beach thousands of miles away from where the animal lived. As a result, although the shells have been known about for hundreds of years, the animal itself was only described in 1832. It was described by the anatomist Richard Owen, best known as the inventor of the word 'dinosaur'.

It is possible that nautilus was known about in ancient times. Around 350 BC, Aristotle wrote a treatise called the *Historia Animalium*, literally 'an inquiry into animals'. In many respects he was the first marine biologist, and this book amply demonstrates his acute powers of observation. He knew, for example, that dolphins and whales were mammals and not fish, a point which even Herman Melville writing *Moby Dick* in 1850, took issue with (although perhaps for literary rather than scientific reasons). Aristotle postulated the first classification of living things, foreshadowing Linnaeus' *Systema Naturae*, while his ideas of adaptation certainly influenced Darwin's thoughts on evolution. In book 4 of the *Historia Animalium*, there is a good account of cephalopods in general, including the first detailed description of their anatomy, but at the end there is a most perplexing passage:

> "There are two others found in shells resembling those of the testaceans. One of them is nicknamed by some persons the nautilus or the pontilus, or by others the "octopus egg"; and the shell of this creature is something like a separate valve of a deep scallop-shell. This octopus lives very often near to the shore, and is apt to be thrown up high and dry on the beach; under these circumstances it is found with its shell detached, and dies by and by on dry land. These octopuses are small, and are

shaped, as regards the form of their bodies, like the bolbidia. There is another octopus that is placed within a shell like a snail; it never comes out of the shell, but lives inside the shell like the snail, and from time to time protrudes its feelers."

ARISTOTLE, *Historia Animalium*, BOOK 4

The above is based on a translation by D'Arcy Wentworth Thompson. His word 'polypus' has been replaced with the modern word for the same thing, 'octopus'. What animal was Aristotle writing about? Is the 'octopus egg' the argonaut? They certainly are weak swimmers, and after storms do get washed ashore. Argonauts are not attached to their 'shells' (the egg case), but hold onto them with their arms; so after death the egg case and octopus can become separated. So far at least the description fits the argonaut nicely, and indeed argonauts are common in the Mediterranean, so Aristotle was probably familiar with them. It is the final line that is problematic. Here 'another octopus' is distinguished from the one that seems to be the argonaut. It is said to be fixed inside its shell like a snail, but capable of extending 'feelers' presumably its arms. Could this line refer to the nautilus rather than the argonaut? It is firmly attached to its shell, and has numerous thread-like arms much more like feelers than the muscular arms of the octopus. These arms do indeed protrude from the shell while the rest of the animal is hidden inside the shell. The similarities between this description and the nautilus are truly striking, but we will probably never know exactly what animal Aristotle was describing. It is difficult to see how he could have ever seen a nautilus: they are not found in the Mediterranean, the Red Sea, or even the western parts of the Indian Ocean.

THE ANATOMY OF *NAUTILUS*

A typical nautilus shell, such as that of *Nautilus pompilius* is smooth, tightly coiled and about 20 cm (8 in) in diameter. The shell is coiled exogastrically, that is, over the 'back' of the animal, like the shells of ammonites but unlike

above: *Nautilus umbilicalis*, a living nautiloid found in the south west Pacific.

those of many coleoids, such as cuttlefish. The outside of the shell is dirty white with brownish-red, irregular bifurcating bands radiating out from the centre of the coil. Overall the pattern is most strongly developed around the upper half of the animal, and virtually absent on the underside. Looking down from above, the dark top of the shell would blend in with the sea floor, while from below the unmarked surface on the underside would match the light filtering down from above. This is called countershading and is common among aquatic animals.

The shell is composed of aragonite, in two distinct layers. The outer layer is made up of vertical prisms, the inner layer of concentric layers of pearly shell. The aperture is sinuous with a notch on either side just below the recessed centre of the shell (umbilicus) where the eye is situated; this is called the optic notch. Further down, there is another notch where the hyponome (or funnel) protrudes. This is the hyponomic sinus. The aperture is closed by a non-calcareous operculum, or lid, made of a leathery substance and attached to the head of the animal. It is sometimes called the 'hood', and is ornamented quite differently from the shell. Whereas the shell sports vivid brown stripes, the operculum is a paler brown with numerous white spots of different sizes randomly scattered over the surface, except for two lines of raised white tubercles on either side of the midline.

If the shell is sectioned through the middle the arrangement of chambers, siphuncle and the main body of the animal are clear. The body sits in the final and biggest chamber in the shell, which is known as the body or living chamber (see plate 1). The animal is attached to the inside of the shell by special muscles, called retractor muscles. Normally the head and arms of the animal protrude from the aperture, but if frightened the muscles pull the animal deeper inside the body chamber and the aperture is sealed off tightly by the hood. If the nautilus is lucky, it will be too difficult for a predator to open up. Behind the living chamber are the smaller chambers which make up the phragmocone, the buoyancy regulating part of the shell, connected to the main body of the animal by the siphuncle.

The soft body of the animal is made up of two main parts: the front part which can protrude from the shell and the back part which is secure inside the shell. The front part includes at least 50 (in females) and as many as 90 (in males) grasping arms surrounding a mouth. The arms are arranged in a circle around a central mouth, or buccal mass. The ends of the arms are somewhat sticky and covered in taste-buds. They are used to find food as well as catch it; in the deep, dark world of the nautilus, the arms are used to touch the sea floor, taste the water, and so turn up the small animals it likes to feed on. These arms are very different to those of the other living cephalopods – they lack suckers or hooks. Instead, each arm consists of a thin, distal part (which is the sticky, taste-bud covered part) and a stout basal sheath into which the distal part can be withdrawn. Curiously, these sheaths are ornamented in the same way as the hood, and it is suggested that the hood and the arms are, in an evolutionary sense, related structures (see plates 2–3).

Like other cephalopods, within the centre of the ring of arms the nautilus is equipped with a sharp beak, although unlike the beaks of coleoids, it has a calcareous tip. The calcified beaks of fossil nautiloids, called rhyncholites, are common, and so this would appear to be typical of the Nautiloidea in general. Behind the beak is the radula. The beak is used to slice up food into manageable pieces which are swallowed whole, while the radula is used to assist in swallowing.

Nautiluses have a pair of 'pinhole camera' eyes, a much less effective kind of eye than the lens eyes which coleoids have. Instead of a lens, a small hole in the front of the eyeball focuses the image on the retina. Unlike a lens, the bigger the hole, the poorer the definition of the image. Pinhole cameras can produce either a dim, sharp image or a bright, blurry one, but it can

above: The Middle Jurassic ammonite *Brasilia bradfordensis* sectioned to show its chambers with calcite crystals in some and hardened mud in others. The body chamber is missing.

never produce a bright and sharp image. So nautiluses must see their underwater world in a fuzzy, vague sort of way. In contrast coleoid eyes can have lenses big enough to make the image bright while still retaining sharpness and contrast.

Much of the back half of the animal consists of a muscular pouch called the mantle cavity. Inside the mantle cavity are the gills, reproductive organs and digestive organs. As with other molluscs, the mantle cavity secretes the shell, but it is also involved in producing the flow of water which brings oxygen to the gills and powers the jet propulsion system. An extension of the mantle, called the hyponome, stretches beyond the aperture of the shell and out under the head. Nautiluses can push water out of the mantle cavity by retracting the head into the shell like a plunger, using special 'retractor' muscles, and the outward flow of water is forced through the hyponome, producing a jet. When the retractor muscles relax, the head springs forwards and water is sucked back into the mantle cavity through the aperture of the shell. By alternating between these two actions, nautiluses generate what is called a ventilation cycle, continually bathing the gills in fresh water from which oxygen can be extracted. Just as importantly, the jet produced can be used for swimming, The retractor muscles are attached to the shell at certain points where they leave distinctive 'scars'. These scars can be seen in some fossil nautiluses and ammonites, indicating that they may have been able to produce a similar ventilation cycle.

As the nautilus grows, it moves itself forward, sealing off part of the old living chamber with a new septum and then filling it with gas. In doing so the extra buoyancy compensates for its extra weight. Living species of nautilus can add a new chamber every month when growing most rapidly, but between the laying down of a new septum, lesser increments of growth leave behind distinctive 'growth lines'. In addition, as the animal grows, it detaches the retractor muscles from their original position and then re-attaches them at a new spot further forward. This leaves a series of scars along the inside of the shell. Fossil nautiluses show similar patterns of growth lines and scars, indicating that they grew the same way.

The shells of nautiluses are often washed up on beaches far from their natural habitats. These 'drift shells' can be easily distinguished from fresh shells since over the months they collect a rich fauna of encrusting animals and plants called epifauna, which are never seen on living nautiluses. These include tube-building worms, bryozoans, barnacles, and algae. Ammonite shells are often found with encrusting animals, particularly oysters, but it is not certain whether these grew when the ammonite was alive, or like the shells of nautiluses, they only became attached after the ammonite was dead.

Chapter Two

AMMONITE FOSSILS

FOSSILS are the remains of organisms that have been preserved in sediments. Some fossils are found in unusual sediments, like the famous tar pits of La Brea in California, or the permafrost of Siberia, but most fossils come from marine limestones, clays and sandstones. Marine environments have high rates of sedimentation, burying corpses quickly, and increasing the chances of fossilisation. Over time, replacement of the original structure of the organism with minerals preserves its overall shape, producing a fossil. Hard tissues, such as bone or shell, have a much higher chance of preservation than soft tissues like muscle or skin, and so the vast majority of fossils are of what palaeontologists call 'hard parts'. In the case of ammonites, the main hard part is the shell, and that is what is most often found.

Ammonite shells are made from aragonite, a mineral that is unstable, and after death it tends to decay. Sometimes it is replaced with other minerals like iron pyrites or calcite, but more often it simply dissolves away. If the shell is filled with sediments before it decays, then the fossil can be preserved in three dimensions, otherwise the fossil is crushed flat leaving an impression in the rock. Some of the most commonly seen modes of preservation are shown on p. 34.

left: Internal cast of an upper Jurassic ammonite showing near empty chambers.

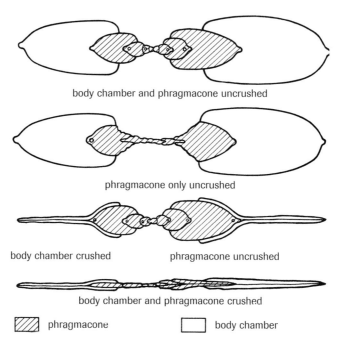

above: Some modes of preservation of ammonites. From top to bottom: a) The body chamber is filled with hardened sediment and the phragmacone (the chambered part of the shell) is filled with crystalline calcite. This is the usual mode of preservation in limestones. b) The body chamber is filled with sediment but the phragmacone is crushed. This sometimes occurs in nodules in calcareous clays. c) The body chamber is crushed flat, but the phragmacone is filled with pyrite and remains uncrushed. This is a common mode of preservation in black pyritous shales and clays. d) The body chamber and the phragmacone are crushed flat. This is the commonest mode of preservation in clays and shales.

AMMONITE SHELL MORPHOLOGY

Ammonite shells are built to the same basic plan as the shells of living nautiluses, but there are significant differences. Both ammonite and nautilus shells are divided into chambers connected by a siphuncle (making up the 'phragmacone'), and both coil exogastrically to form a regular, flat spiral. However, ammonite shells differ significantly in the shape of the septa dividing the shell into chambers, the position of the siphuncle, and the shape of the embryonic shell.

SEPTA

The septa dividing the shell of any cephalopod into chambers are shallow concave walls, but what distinguishes ammonites from nautiluses and coleoids is that the septa are thrown into folds radiating outwards from the centre and becoming more tightly folded as they merge with the wall of the shell. The pattern the septum makes at its junction, or suture, with the wall is called the suture line. The suture line varies greatly among different groups of ammonites, and so has become very important in the classification of ammonites, as is described in Chapter 5.

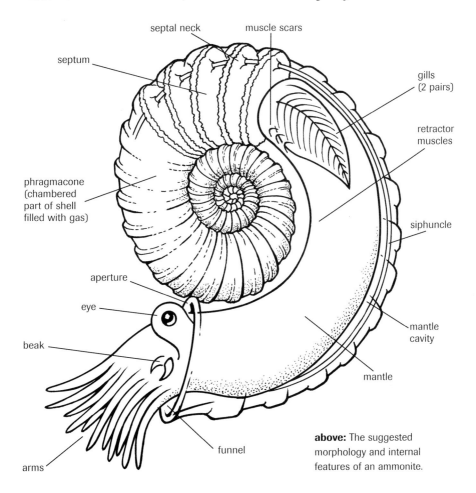

above: The suggested morphology and internal features of an ammonite.

Siphuncle

In nautiluses, the siphuncle runs more or less through the centre of the septa, and where the siphuncle pierces a septum, a short 'collar' surrounds the siphuncle, called the septal neck. This neck extends backwards from the septum into the preceding chamber. Such a backwards-pointing (or retrochoanitic), septal neck is common in fossil nautiluses as well as in living species of nautilus. In contrast, the septal necks of most ammonites point forwards, and are termed prochoanitic. Prochoanitic septal necks are also seen in those coleoids which have chambered shells, such as belemnites, and so is yet another character that ammonites share with coleoids.

Embryonic Shell

One final difference between ammonite and nautilus shells is the shape of the embryonic shell, the part of the shell which developed before the animal hatched from its egg. In living nautiluses, the eggs are large for a cephalopod, and inside the egg the nautilus embryo develops a cap-shaped shell made of an organic material called chitin. In contrast, ammonite eggs are small, about the same size of those of squids, and the embryos secrete a calcareous barrel-shaped shell very similar to those of embryonic coleoids such as belemnites. This initial shell, the protoconch, is made from aragonite. The growth and development of ammonites is described in Chapter 4.

Coiling and Ornamentation

Ammonites exhibit a wide range of coiling modes. Some resemble nautiluses in being involute, i.e. the whorl section is crescent or kidney-shaped, and later whorls wrap over the earlier whorls, hiding them from view. Involute forms include oxycones, where the shell is disc-like as it is highly compressed from side to side (see plates 6–8), and cadicones, which are compressed from top to bottom, producing a globular shell. Other

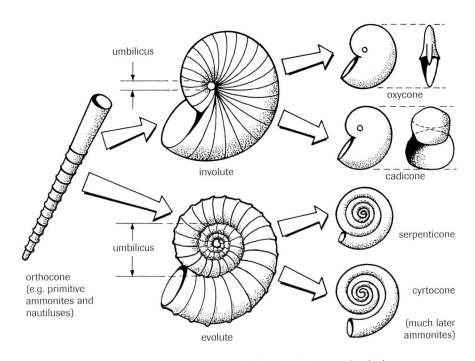

above: Common types of ammonite shells. Only primitive ammonites had orthocone shells, most others were coiled spirals.

ammonites are evolute, with later whorls only partly obscuring the earlier whorls, which can be clearly seen at the umbilicus. Extreme forms of evolute coiling include serpenticones, where later whorls just touch the earlier ones, and cyrtocones, where successive whorls are not in contact at all. Further variations in coiling mode can be seen among the heteromorphs, which begin as spirals but then unfold in various ways. Some are straight, some are helical and some are a combination of coils, straight sections and hooks. The heteromorph ammonites are discussed in Chapter 5.

The opening of the shell, the aperture, is in some species plain and similar to those of modern nautiluses, but in many ammonites it is much more

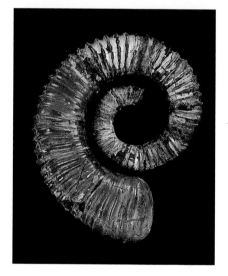

top left: An oxycone ammonite, *Amaltheus* sp.
top right: The Lower Jurassic evolute ammonite *Asteroceras stellare* which is openly coiled.
bottom left: The Upper Jurassic serpenticone ammonite *Dactylioceras commune* showing whorls just resting on previous whorls without overlap.
bottom right: The Lower Cretaceous crytocone ammonite *Aegocrioceras quadratus* in which whorls do not touch each other but form an open spiral.

elaborate. Extensions of the shell from the sides of the aperture, called lappets, and from the bottom, called rostra, are particularly common in Jurassic and Cretaceous ammonites.

A few ammonites have smooth shells, like those of nautiluses. However, most have ornamented shells, with ribs, keels and spines. These are modifications of the shell produced by the mantle as the animal grew. The ribs may be fine or bold, simple or forked. In some species ribs run uninterrupted from the recessed centre of the shell (umbilicus) on one side of the shell over the ventral surface and up to the umbilicus on the other side. In other species longitudinal ridges, called keels, cut across the ribs along the ventral surface. Spines are not common in Palaeozoic or Triassic ammonites, but they became increasingly widespread and well-developed

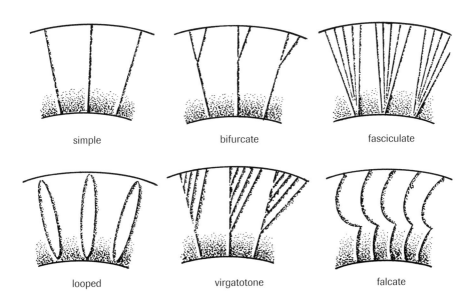

above: Six of the commonest types of ribbing found on ammonites. Simple ribbing characterises early Jurassic forms while virgatotome ribbing occurs on late Jurassic forms.

above: The Lower Jurassic ammonite *Amaltheus gloriosus* in which the ribs are raised to form a keel seen laterally (left) and ventrally (right).

in Jurassic and Cretaceous ammonites (see plates 9–10). They are believed to have evolved as a defence against predators, but how they were formed is not clear. The spines are usually hollow, and so presumably extensions of the mantle protruded from the edge of the aperture to secrete them. As the animal grew larger and added new chambers, the spines that were once at the edge of the aperture wound up further and further back, eventually on the chambered part of the shell. Being hollow, if these broke off they would release the gas from inside the chambers, reducing the animal's buoyancy. To prevent this the spines were sealed off after they were formed. Indeed, it is common to find ammonites with the spines broken off, and the sealed-off stumps left in place. These stumps are known as tubercles or spine-bases.

AMMONITE FOSSILS | 41

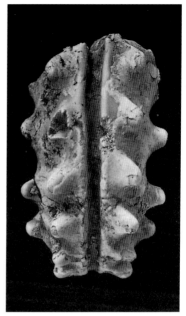

above: The Lower Cretaceous ammonite *Euhoplites armatus*. Two lines of stumps are visible which were the bases of spines that have since been lost.

The coloration of ammonites is only hinted at in the fossil record, as pigment is very rarely preserved. Traces of longitudinal stripes have been seen on some ammonites (see plates 11–12); these are more like garden snails than the tiger stripe pattern of the modern nautilus.

THE SUTURE LINE

The figure on p. 42 shows a specimen of *Lytoceras* which has been preserved as an internal mould, that is, sediment has filled the hollow chambers of the shell, and the shell has dissolved away. The septa have gone, but left behind as a suture between the moulds of each chamber is a tracery of what the junction between the septum and the wall of the shell looked like, called the suture line. Another way of looking at the shape of the septa is too look at the internal mould of a single chamber, as with the

42 | AMMONITES

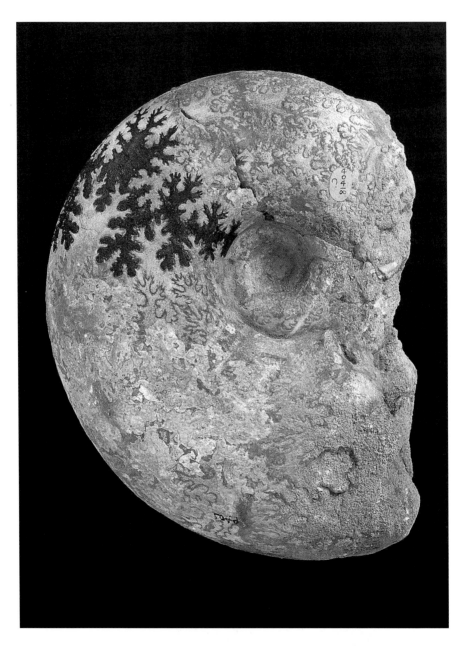

above: This specimen of *Lytoceras aaleniarum* has had one of the chambers painted black to make it easier to see the suture line.

specimen from *Euaspidoceras* below. The front view shows four extensions reaching forwards, called 'saddles'. In between the saddles are inwards pointing folds called 'lobes'.

As mentioned earlier, the shape of the suture line and the number of lobes and saddles are very useful in classifying ammonites. Most ammonite species have a distinctive suture line, which can be used for identification, like a sort of ammonite fingerprint. For example, *Hamites hybridus* has the four-lobed pattern typical of heteromorphs on each side of the shell: the external, lateral, umbilical and internal lobes (see p. 44). These may be bifid, with two approximately similar 'leaves' to the lobe, or trifid, with three 'leaves' one on each side of the line of symmetry, and one in the middle of the line of symmetry. The siphuncle usually lies close to the external lobe, on the outside of the whorl. Suture lines are especially useful in the identification of an ammonite where the rest of the fossil is not well preserved.

left: The back of a septum in *Euaspidoceras* in which the ventral, dorsal and lateral lobes form a vertical cross. The front view shows four saddles forming a diagonal cross.

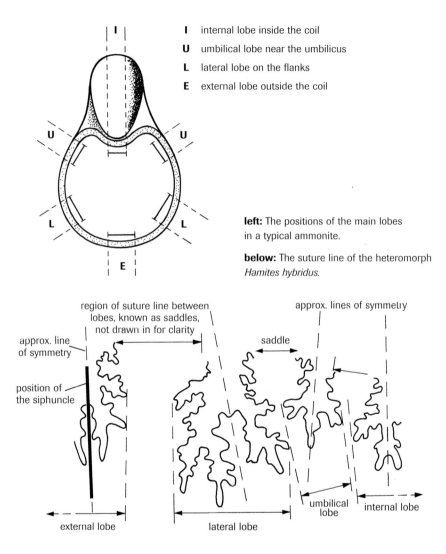

left: The positions of the main lobes in a typical ammonite.

below: The suture line of the heteromorph *Hamites hybridus*.

In general, primitive ammonite groups have relatively fewer folds than more advanced ones. The very earliest ammonites, such as the Anarcestina, have suture lines that are barely distinguishable from those of nautiluses. Later groups, like the Goniatitina and Ceratitina have successively more folds, while the suture lines of the Ammonitina are the most complicated of all, as is described further in Chapter 5.

THE APTYCHUS

Occasionally an ammonite is found with a shield-like structure in the body chamber. In fact there are two types: the aptychus and the anaptychus. The aptychus (plural aptychi) is composed of two parts, each a mirror image of the other, often joined along a common straight edge. The anaptychus (plural anaptychi) is a single plate, preserved as a carbonaceous film on an impression, indicating a chitinous composition, but sometimes with a thin shelly film covering the surface.

Aptychi are made of calcite, unlike the shell of the ammonite which is of aragonite, and therefore the aptychus may not have been formed by the mantle, which laid down the shell. Aptychi are found in such common ammonites as *Hildoceras*, *Haploceras*, *Phylloceras*, *Stephanoceras*, *Perisphinctes*, *Scaphites*, and *Aspidoceras*. While they are not known any earlier than the early Jurassic, none are known after the end of the Cretaceous. This is when the ammonites died out, which seems to support the view that aptychi are specific to ammonites. Anaptychi are known from much older rocks, as old as late Devonian in age, and have been found in ammonites such as *Eoasianites*, *Psiloceras*, *Eoderoceras* and *Lytoceras*.

Sometimes the individual parts of aptychi are found without being associated with an ammonite, and since the plates of aptychi are very bivalve-like, with concentric ridges around a point of origin, they were first described as bivalves. Only when they were found in the body chamber of an ammonite was it realised that they were part of the ammonite. It was thought that, by analogy with the operculum of certain marine snails, the aptychi probably served the same function, namely to close the aperture of the shell against predatory intruders. Then came the doubts, founded on the observation that, in some cases the presumed 'operculum' did not fit the aperture of the shell precisely and therefore could not function in that way. It was in 1864 that two Americans, Meek and Hayden, discovered a

Scaphites from the Upper Cretaceous of Dakota with the aptychus in the body chamber but with the clear impression of the upper part of the jaw in between the two parts of the aptychus. After much thought they stated that '... we must then view the two enveloping valves as forming together one of the opposing mandibles'.

In the 1930s, Trauth, a Viennese palaeontologist, published a number of papers on aptychi, defending the view that they were used by the ammonite

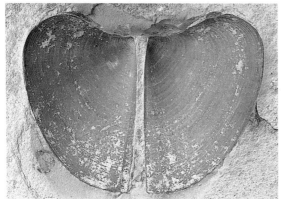

above: The aptychus in place at an aperture.

left: The two parts of an aptychus, separated from an Upper Jurassic ammonite, and lying loose in the rock.

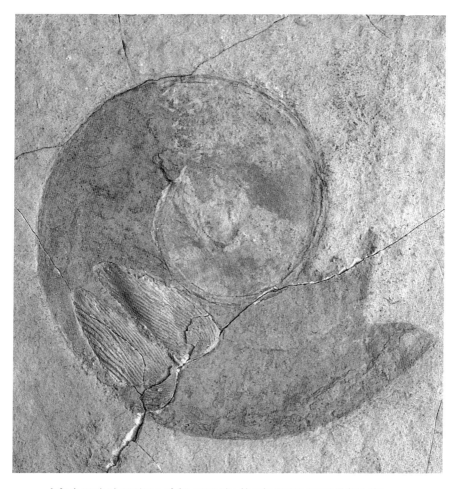

left: A crushed specimen of the ammonite *Neochetoceras steraspis* from the Upper Jurassic of Germany, with the aptychus in place in the body chamber.

to close off its shell, and his view was supported by the work of Schindewolf during the 1950s, who published a photograph of an Upper Jurassic ammonite, *Physodoceras*, from Germany which had the aptychus in place at the aperture of the ammonite shell. But by the 1970s, ammonite researchers like Lehman had returned to the jaw theory. For example the body chamber of a Lower Jurassic *Pleuroceras* specimen contained a structure which Lehman interpreted as an upper jaw lying between the folded edges of an anaptychus.

Lehman makes an interesting argument which is cogent and supported by evidence, but it is counter-intuitive. Consequently, there remain many who doubt his interpretation, partly on the grounds that jaws of living cephalopods are, relative to the size of the body, very small. The giant squid *Architeuthis* is about 10 m (33 ft) from the tip of its tentacles to the end of its tail, but has a jaw apparatus less than 20 cm (8 in) across (about one-fiftieth of its body length). The jaws of the 30 cm (12 in) long common squid *Loligo forbesi* are about 1 cm (0.5 in) or so across, and so proportionally similar to those of its giant relative. But an aptychus from a typical ammonite such as the specimen of *Neochetoceras* is one-fifth of its body length. In other words, if the aptychus was part of the ammonite's jaw, then ammonites were radically different from typical cephalopods in this respect, with simply enormous jaws. The matter is still hotly debated, and the factual evidence of Lehman and Schindewolf seems to favour both sides of the argument.

AMMONITE SOFT BODY ANATOMY

Most of what we know about the soft body parts of ammonites is based on inductive argument by analogy with living cephalopods. There are certain features present in all living cephalopods which we may safely assume to have been present in ammonites. A constant feature is a head with grasping arms surrounding a mouth, within which is the formidable beak, and behind the beak is the radula. There are eight arms of equal length on an octopus, while squids and cuttlefish have ten, of which two, the tentacles, are much longer than the others. The nautilus has between 50 and 90 arms, but they are much finer and without suckers. We do not know if ammonites had eight, ten, or 90 arms, whether they had suckers on the arms or if they were plain like those of the nautilus.

Like other cephalopods, ammonites were probably carnivores, and the few fossils found with remains of the gut seem to bear this out. Inside such fossils, fragments of small crustaceans and echinoderms, plankton, and even smaller ammonites have been found. Carnivores generally have a short gut with a simple stomach, since meat is much easier and quicker to digest than plant material. In squids and cuttlefish, there is a large digestive organ, or liver, connected to the gut, and when fully digested the inedible remains empty into the dorsal side of the mantle cavity via the anus.

The mantle cavity also contained the gills, the structures which took up oxygen from the surrounding water and got rid of waste carbon dioxide. The nautilus has four gills, two on each side of the animal, while coleoids have just two. Traditionally, the ammonites have been assumed to have had four as well, simply because this was thought to be the primitive condition,

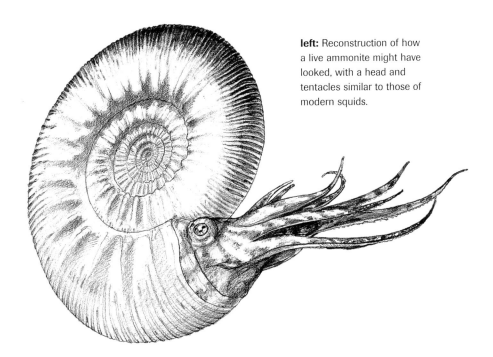

left: Reconstruction of how a live ammonite might have looked, with a head and tentacles similar to those of modern squids.

and ammonites must have been primitive cephalopods like nautiluses. Now that ammonites are seen to share so many other features with coleoids, this view has been dropped, and two gills is thought more likely. But again, until a fossil shows them in place, this is just guesswork. The flow of water inside the mantle cavity was probably similar to that of other cephalopods, particularly the nautiluses, as we have described earlier. Water would have entered the mantle cavity, passed over the gills, the oxygen extracted, and then forcibly ejected through the hyponome carrying the animal's waste away with it.

The bloodstream of modern cephalopods is closed, that is, the organs and tissues are supplied with oxygen by blood which travels in veins and arteries, and is pushed around by the pumping action of a heart. In fact,

above: *Sepia officinalis,* common cuttlefish. Note the camouflage 'tiger stripes' on the dorsal (top) of the animal, and the enlarged lowest arm.

cephalopods have a number of 'hearts', one for each gill, and one more to send the oxygenated blood around the body, although their blood pressure is very low compared to that of a comparably sized vertebrate such as a fish. Ammonites were presumably no different. A peculiarity of cephalopod blood is that it is based on a copper pigment called haemocyanin. Haemocyanin is the most common respiratory pigment seen in molluscs, but is much less efficient at carrying oxygen than haemoglobin, the pigment in vertebrate blood. Although inferior to haemoglobin, haemocyanin works well enough for the low level of activity shown by most molluscs, but their relatively poor ability to transport oxygen around their bodies has been an evolutionary handicap to cephalopods. Compared with similar sized fish, cephalopods have much less stamina. For example, squids can sprint as well as any fish, but over long distances they tire more quickly.

All the soft body parts of the ammonite would have been enclosed by the mantle, and attached to the shell by the muscles. Since the posterior end of the animal must have formed each septum as the animal grew in width and moved forward into its enlarging shell, it must have taken on the shape of the septum before beginning to form it. It should not be forgotten that the mantle would also need to surround part of the siphuncle to produce the septal neck.

Chapter Three

AMMONITE FORM AND FUNCTION

MOST molluscs have shells, and cephalopods are no exception. What distinguishes them from other molluscs is the way their shells work. Gastropods and bivalves have shells primarily to protect them against predators, but thick, strong shells are heavy, and this prevents them from being fast-moving animals. In contrast the shells of cephalopods, while retaining their strength, are not heavy because of the gas-filled chambers, the phragmocone. It was the acquisition of this neutrally buoyant shell combined with the ability to swim using jet propulsion that made the cephalopods such successful marine predators during the early Palaeozoic. To some extent they have retained this advantage over invertebrates, although vertebrates, particularly bony fish, have become pre-eminent in modern oceans.

There are four key elements to understanding cephalopod form and function – neutral buoyancy, orientation, jet propulsion, and streamlining – each of which places compromises on the shell's defensive strength. Ultimately, cephalopods have had to trade-off increased mobility and swimming efficiency against the defensive properties of the shell. The shell can be thought of as being under a constant 'crushing attack' from the surrounding

left: A specimen containing *Asteroceras* (large shells) and *Promicroceras* (small shells) from near Yeovil, Somerset, UK.

water. The thinness of the shell and the lack of structural support between each septum, means the shell has little extra strength to resist the additional forces from a predator's bite. The shell could be thicker, but only by losing some of the empty space inside, and so with a loss of buoyancy. In short, the shell can be thick and strong but heavy, or thin and weak but buoyant, but not both. This is a perfect example of what are called biological compromises, the end result of evolution having to deal with contradictory but essential demands. Understanding that such compromises exist undermines the ancient belief in organisms being in some way 'perfectly designed' for their way of life, and that every aspect of their anatomy is fitted to their ecology, a belief known as teleology. Evolution is instead all about compromises, and advantages are almost always initially only marginal, but they are cumulative, and the vast periods of geological time allow these to build up into the amazing diversity of life we see today.

In Chapter 1 the appearance of the first cephalopods in the early Cambrian was described. At that time, most of the other invertebrates, such as trilobites and gastropods, lived on the sea floor. Swimming above the sea floor were jellyfish, a few small trilobites and some primitive fish-like chordates, the ancestors of the vertebrates. None of these swimming animals were powerful swimmers, and they either kept close to the sea floor or drifted with the plankton. As a result, most animal life proceeded at a somewhat pedestrian pace: either crawling over the sea floor or burrowing through it, or staying still and waiting for prey to wander within reach. Predators and prey lived their lives on the sea floor, and that was where they kept their eyes and antennae focussed.

The appearance of the cephalopods shattered the ambulatory and two dimensional world of the late Cambrian. Cephalopods were able to cruise along almost effortlessly, and so could search large areas for prey quickly.

Like owls snatching mice, death came from above for the bottom living invertebrates of the Cambrian period, and it came rapidly and without warning. Of course it was only a matter of time before life adapted to these new predators, some by developing stronger shells, others by learning to burrow into the sand or otherwise make themselves inconspicuous, but for a while these cephalopods were simply the most dangerous animals around.

Modern nautiluses retain the chambered shell as a flotation device, but only a few coleoids do, for example the cuttlefish and a deep-sea squid-like animal called *Spirula*. Although squids do not have chambered shells, some have evolved an alternative method of acquiring neutral buoyancy – the use of lightweight chemicals in their tissues, typically ammonium ions (NH_4^+). Ammonites not only retained the chambered shell, but refined it to such a degree that many ammonite shells seem over-complicated. To understand why ammonites had such a diverse array of shell shapes, we must first understand how they worked in relation to buoyancy, water pressure, orientation, and streamlining.

BUOYANCY

In the third century BC, the king of Syracuse (modern-day Sicily), Hieron II, commissioned a goldsmith to produce a new crown to be used in an upcoming religious festival. Hieron gave the artist a certain amount of gold, but he was convinced that the craftsman had stolen some of the gold, and that the finished crown had been adulterated with some base metal. But how could he tell? It weighed just as much as the quantity of gold Hieron had provided to make the crown. So the king asked Archimedes to solve his problem.

As everyone knows, the solution came to Archimedes whilst he was taking a bath. Archimedes noticed that the surface of the water in the bath rose when he got in. It is certain that people had overfilled their baths before, but Archimedes was the first to connect the amount of water that spilled out of the bath with the volume of his body. This connection now bears his name, Archimedes' Principle, and it states that the volume of water displaced is equal to the volume of the body placed into the vessel containing the water. Note that the weight of the object, or the substance it is made of, is irrelevant. Archimedes would displace the same amount of water whether he was made of flesh and bone, bronze or wood. It was this fact that revealed how Archimedes could solve King Hieron's problem, and supposedly he was so excited by his discovery that he ran straight to the palace, without stopping to put on his clothes, shouting "Eureka! Eureka! – I've got it! I've got it!"

History does not record exactly how Archimedes employed his observations to test the quality of the crown. Most probably he would have called for an amount of pure gold similar to that the king had given the goldsmith. Placed into a vessel of water he could then measure how much water it displaced. This quantity would be proportional to the volume of that lump of gold, not its shape or size. He would then repeat the experiment using the crown, and find out its displacement. If the crown was pure gold, then the amount of water it displaced would be identical to that of the lump of pure gold. Unfortunately for the goldsmith, the crown displaced too much water. This could only mean one thing: it had a greater volume than the lump of gold, so could not be pure gold. The goldsmith had replaced some of the gold with another metal, and so attempted to defraud the king. The fate of the goldsmith is unknown, but doubtless it was unpleasant.

Archimedes' Principle operates on the relationship between the volume of an object and the volume of water it displaces. But it leads to a further development, the Principle of Buoyancy – why some objects sink and some float. This principle deals with the ratio of the volume of a body and its mass (density). Mass is the amount of matter in a body, and weight is the measure of mass. If the mass/weight of a body is high and the volume is low, it will sink in water. Reverse the ratio, high volume and low mass/weight, and the body will float in water as seen in this example.

Imagine a submerged submarine which in air weighs a thousand tons. The submarine has a fixed volume, but it can vary its mass by letting water into its buoyancy tanks, or pushing it out with compressed air. If the thousand-ton submarine displaces exactly a thousand tons of water then it will neither sink nor rise – it is neutrally buoyant.

By letting water into the tanks, the submarine weighs more (increases its mass) and it sinks. Forcing water out with compressed air has the reverse effect, the submarine weighs less (decreases its mass) and rises. Living cuttles, the odd little *Spirula*, and nautiluses use their phragmacones ('buoyancy' tanks) in this way, and it is reasonable to suppose that the phragmacones of fossil nautiluses, belemnites and ammonites functioned in this way.

Cephalopod tissues are slightly more dense than sea water, and the material from which the shell is made, aragonite, is almost three times more dense than sea water. This means that the cephalopod tissues and shell weigh more than the water they displace, and so the upwards force is not enough to make them float. Adding chambers to a shell and then filling them with gas increases the volume of the animal without adding significantly to its weight. If the volume of water displaced is sufficiently great that the lift produced matches the weight of the animal, then it will neither float nor sink, but float in mid water. This is precisely what modern nautiluses and cuttlefish do, and presumably so did ammonites, allowing them to hold their position in the water column without actively swimming. In contrast, squids, lacking a chambered shell, usually weigh more than the water they displace, so tend to sink, and must swim continuously to stop sinking.

As described earlier, nautiluses build their chambered shells throughout their lives, adding new chambers as they grow. It is the siphuncle that does the work of emptying the chambers. The siphuncle has a semi-permeable membrane, which means that it only allows small molecules, like gases and water, to cross it. Larger molecules, like salts, sugars and proteins, cannot. When a chamber is completed, it is at first filled with ordinary sea water. The nautilus then modifies the acidity and salt content of the blood which flows through the siphuncle. The increase in salt sets up an osmotic gradient between the bloodstream and the sea water in the chamber. The water then moves from the less concentrated sea water into the more concentrated blood. Because the siphuncle is only semi-permeable, salt cannot escape from the blood into the sea water. The saltiness of the blood in the siphuncle does not decrease, and so water keeps moving from the chamber into the siphuncle.

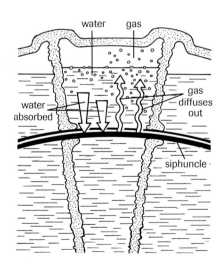

above: Over time the siphuncle draws out the water 'emptying' the chamber, while gas diffuses out of the blood in the siphuncle filling the empty space.

The net effect of this is to reduce the amount of water in the chamber. This reduces the pressure in the chamber as the water level drops, which causes gas to diffuse out from the blood and into the chamber. The change in the acidity of the blood further promotes this by making gases less soluble in the blood. The overall process is therefore one of draining water out of the chamber and letting gases passively diffuse in. In this way the nautilus produces a shell which is large but very light, reducing its overall density and giving it neutral buoyancy. Of course, as the nautilus grows in size it must compensate for its increase in weight, but there are otherwise no limitations on the potential size of the animal provided it can keep adding and emptying new chambers. Besides the nautiluses, the few coleoids that have chambered shells employ exactly the same system, and ammonites are believed to have done so as well.

WATER PRESSURE

Water pressure on an object increases the deeper you go, and for a hollow object like a gas-filled chambered shell it is a major problem. Even on land, we are exposed to the weight of the 500 km (310 miles) thick atmosphere pressing against our bodies. This force, called air pressure, is at sea level said to be '1 atmosphere'. This translates into a force equivalent to a weight of 1 kg pushing down on an area of 1 cm^2 (15 lb/in^2). We do not notice this force because our bodies are adapted to cope with it. This sounds like a lot of force, but bones, tissues and fluid-filled cells resist compressive forces well, and this amount of pressure does no harm. Furthermore, air pressure changes very little. Moving up or down tens or even hundreds of metres makes very little difference to the pressure to which an animal is exposed. Even a visit to a high altitude city like Denver, Colorado at an altitude of 3000 m (9843 ft) subjects the visitor to an air pressure only 2 or 3 per cent less than at a sea level city like New York. The air is said to be a little 'thinner', meaning a given volume of air has less oxygen than it would at sea level. Since our lungs do not change in size as we travel around, a breath-full at Denver holds less oxygen than it would at sea level, and some people feel a bit faint for the first few days until they get used to it.

Water is a much more rigorous medium than air in this respect, since it is much more dense, and much heavier. At the surface of the sea, water pressure is exactly the same as it is on dry land at sea level, namely 1 atmosphere, but it increases by 1 atmosphere with every additional 10 m

(33 ft) depth. Solid objects are not greatly affected by water pressure, since they are non-compressible, and rocks and stones on the bottom of the sea have the same shape and volume as they would on dry land. The same holds true for water, which is also virtually non-compressible, and 1 kg (15 lb) of sea water has the same volume at the bottom of the ocean as it does at the surface. Hollow objects are very different, as gases compress readily with increasing depth. Manufactured structures tend to be hollow and filled with air, so are very sensitive to water pressure. Submarines are a good example of such a structure, being essentially hollow tubes filled with air. Anyone who has seen movies about submarines will be familiar with unsettling creaks as the metal strains to resist the crushing force of the surrounding water. Even with the best design and strongest metals, submarines are only able to go down to relatively trivial depths compared with the greatest depths of the ocean. An 80 m (262 ft) depth is the sort of dive a submarine might do to try and escape from a destroyer, whereas most of the floor of the ocean is between 1000 and 5000 m (3280 and 16,400 ft) further down. Even at 80 m (262 ft), a submarine has to resist a water pressure of no less than nine times atmospheric pressure. Implosion – the catastrophic failure of the hull to resist the force of water pressure – is a very real danger to submarines. At 200 m (650 ft), a submarine must resist 21 atmospheres, which is roughly equivalent to 210 tonnes (230 tons) of weight being pressed down on every square metre of the surface of the ship! This is a staggering amount of force, and is about the limit for most military submarines. Only special deep-sea submersibles can go any further down, and they are all very small vessels with thick hulls and are usually spherical in shape.

When oceanographers started diving to the bottom of the sea in submersibles they found to their astonishment that there were lots of animals living there. Even the appropriately named abyssal plain teemed with life. The abyssal plain lies (on average) below 4000 m (13,124 ft) of water, and so the ambient water pressure is over 400 atmospheres, which translates to about 4000 tonnes/m^2 (4400 tons/in^2). In this harsh and sunless realm can be found thousands of different species of anemones, corals, worms, snails, sea cucumbers, starfish, shrimps, crabs, and fish. Some have shells, but many, like the anemones and sea cucumbers, are completely soft bodied and often rather watery or flabby in appearance. Provided the shells and bones are solid and any soft tissues are completely filled with water or some other liquid, then the pressure does not crush these animals in the way it would a submarine. While there are numerous cephalopods at this depth, mostly squids and octopuses, there are no cephalopods with chambered shells. Cuttlefish are only found in water of less than 200 m (650 ft) depth, and the nautiluses between 200 and 400 m (650 and 1312 ft). The deepest diving cephalopod with a chambered shell is *Spirula*, which inhabits waters down to about 1000 m (3280 ft).

Why are cephalopods with chambered shells absent from the deepest parts of the ocean? Nautiluses have been intensively studied in this regard, and the limitation appears to be water pressure. Below a certain depth, the shell is unable to resist the compressive forces of water pressure. This limit is at around 700 m (2296 ft), and beneath that depth the shell is at risk of imploding, which is fatal for the animal. Cuttlefish are even more severely limited. They have a very distinctive shell, the cuttlebone, which is divided into thousands of tiny thin-walled chambers, giving it a rather flaky texture. This design allows the animal to change its density very quickly, and unlike the nautilus can actually use these changes to rise or sink in the water column like a submarine. However, this open design makes the shell structurally weak, and this restricts cuttlefish to shallow water. The only cephalopod with a chambered shell living in deep water, *Spirula*, seems to lessen the problem of water pressure by being very small. Its shell is tiny for a cephalopod, only 2–3 cm (1 in) in diameter, and its walls are proportionately thicker.

Ammonite shells are chambered like those of the nautilus, so the assumption is that the ammonites lived at a similar range of depths. However, as we have seen, the septa of ammonites are far more complex than those of the nautilus. Many palaeontologists believe that the elaboration of the junction between the septa and the wall of the shell was an adaptation to living in deep water where the water pressure is greater. The folds could have spread out the forces produced by water pressure more evenly around the shell, reducing the stress on the junction between the septum and the shell wall. However, ammonites are not commonly found in deep water sediments, and the majority seem to have inhabited shallow coastal seas between about 50 and 200 m (165 and 650 ft). So why did they need to be adapted to deep water, when the modern nautiluses, which live in much deeper water, have shells with very simple sutures. One possible solution to this contradiction is that ammonites did not go into particularly deep water, but by having shell designs that were intrinsically more robust they could make them thinner and lighter than those of nautiluses. In this way they would have been able to produce shells more quickly using less material, and consequently grown much faster. An alternative explanation of the function of the complex septum implies they were an adaptation not to water pressure but to predation. The suture line is seen as a mechanism for distributing what are called 'point forces' around the shell. Point forces are produced when sharp or pointed objects, such as the teeth of a shark, press against the shell. The force is concentrated in a small area, the point, and if the force is sufficient, the shell is pierced. If a point force can be distributed around the shell, the stress at any one point is reduced, and it becomes less likely that the shell will break.

ORIENTATION

It should now be clear how ammonites were able to attain neutral buoyancy, within the limits set by water pressure. However, simply being

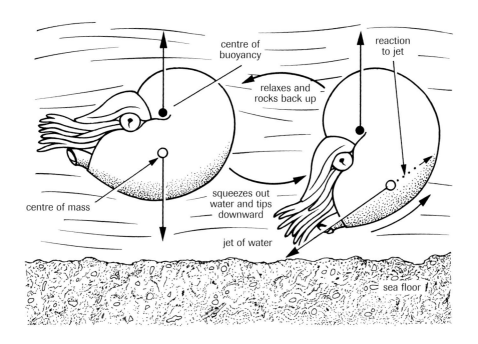

above: The centre of mass is vertically below the centre of buoyancy so that when the nautilus pumps water out of the hyponome, it causes the shell to rock, like pushing a pendulum. This uses energy which is not used for propulsion and is therefore an inefficient mode of swimming.

able to float is useless if the animal cannot swim properly, catch its food, or see its enemies. In other words, it must be oriented properly. The orientation of a floating object depends on the relative positions of the centre of mass and the centre of buoyancy. Normally an object will float with the centre of mass vertically below the centre of buoyancy.

It is easy to imagine this if one considers a hot-air balloon. The basket in which the passengers stand – the gondola – is much the heaviest part of the whole structure, so the centre of mass of a hot-air balloon can be assumed to be in the gondola. On the other hand, the volume of the gondola is so small compared with the huge bag which contains the hot air

– the envelope – that the centre of mass of the air displaced by the balloon will be somewhere inside the envelope. Consequently, the balloon will float with the gondola directly below the envelope. If the gondola is pushed by a gust of wind to one side, then the centre of mass is no longer directly below the centre of buoyancy. This is an unstable orientation, and the gondola will swing back until it is directly below the envelope once more.

Cephalopod shells are like underwater balloons. The animal itself is small but heavy, and occupies only the final chamber of the shell. This is where the centre of mass is. The chambered part of the shell, the phragmocone, is light but accounts for most of the volume of the animal, and so displaces most of the water. The centre of buoyancy is here. When free-floating, the orientation of a cephalopod such as a nautilus will be with the centre of mass, the body, vertically below the centre of buoyancy, the chambered shell. For coiled nautiluses and ammonites this means that the body chamber is close to the sea floor, and that the aperture points forwards and slightly upwards.

right: In many species of ammonites, the aperture appears to have pointed more or less upwards.

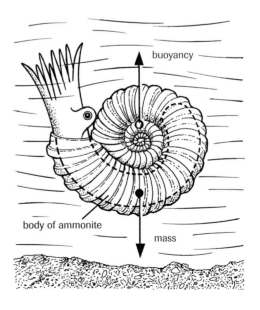

plate 1: The shell of the nautiloid *Nautilus pompilius* sectioned to show body chamber, septa and siphuncle.

66 | AMMONITES

plate 2: Living *Nautilus pompilius*
It shows the many arms, the hood,
and the large 'pin-hole' eye.

plate 3: Front view of
Nautilus pompilius, showing
the overlapping folds
of the hyponome.

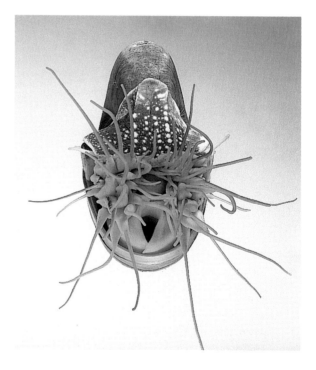

AMMONITES | 67

plate 4: *Ichthyosaurus grendelius* with young, eating an ammonite.

plate 5: An ammonite floats above crinoids, sea urchins, brachiopods and molluscs.

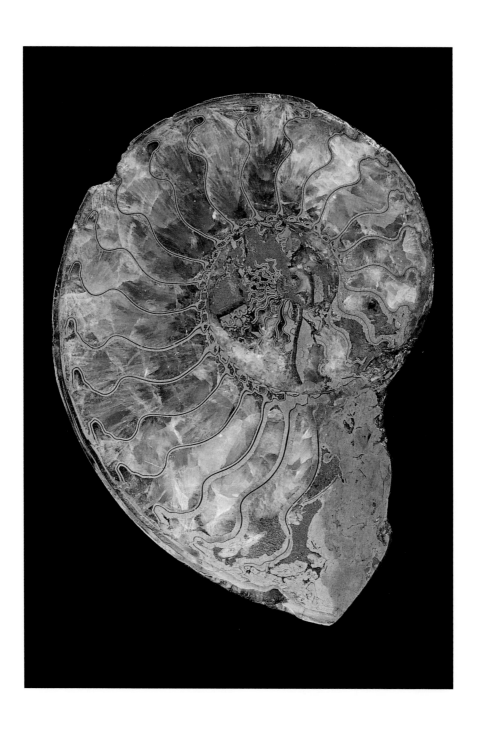

AMMONITES | 69

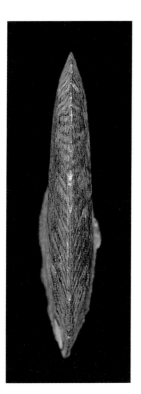

plates 6, 7, and 8: The Lower Jurassic oxycone ammonite *Oxynoticeras oxynotum.*

The sectioned photograph (opposite) shows tall chambers with green calcite and yellow pyritised septa; above left is a ventral view, and above right is a lateral view of the same specimen.

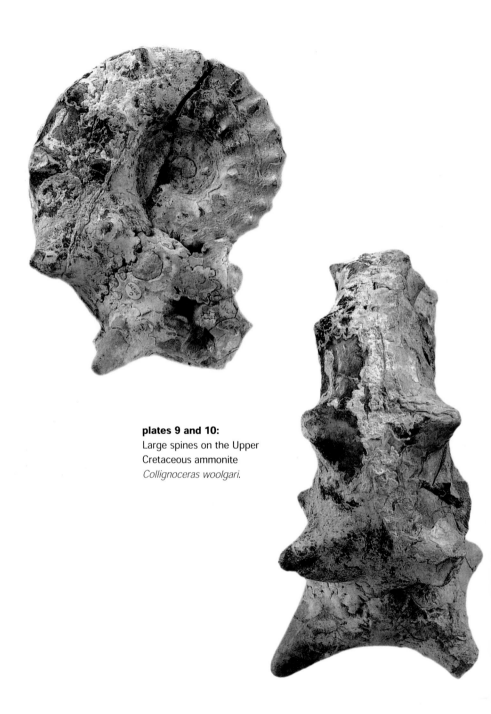

plates 9 and 10:
Large spines on the Upper Cretaceous ammonite *Collignoceras woolgari*.

AMMONITES | 71

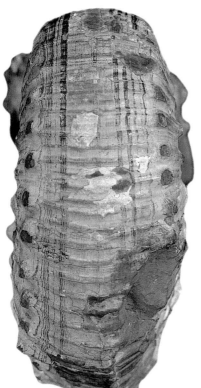

plate 11: Spiral lines on the phragmacone and body chamber of the Lower Jurassic ammonite *Liparoceras naptonensis*. These lines break up the outline of the animal in order to confuse a predator.

plate 12: A specimen of the fossil nautiloid *Aturia* from the Miocene of Australia, which, despite being 30 Ma old, preserves some traces of its 'tiger stripe' coloration, very similar to that of living species of nautilus.

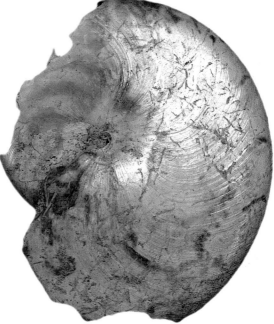

plate 13: Microconch with lappets, of the Upper Jurassic ammonite *Aulacostephanus autissiodarensis*.

plate 14: Macroconch of the Upper Jurassic ammonite *Aulacostephanus autissiodarensis*.

plate 15: The partially uncoiled Upper Cretaceous ammonite *Scaphites nodosus* from the USA.

AMMONITES | 75

plate 16: *Bostrychoceras polyplocuin* from the Upper Cretaceous of Germany. The coiling is loose and slightly open, and the aperture is on the right hand side (dextral) instead of the usual left hand side (sinistral).

plate 19: Part of the phragmacone of an Upper Cretaceous straight ammonite, *Baculites* sp.

AMMONITES | 77

plate 17: Middle Jurassic *Parkinsonia dorsetensis* ammonite has been polished to show 'suture' lines, frilled edges of the septa which separate the chambers.

AMMONITES

plate 18: *Titanites* sp. a macroconch from the Upper Jurassic of the UK.

AMMONITES | 79

plate 20: *Scaphites* sp. preserved with some of the original shell, the metalic white 'mother of pearl' layer. Normally, the actual shell is replaced with minerals.

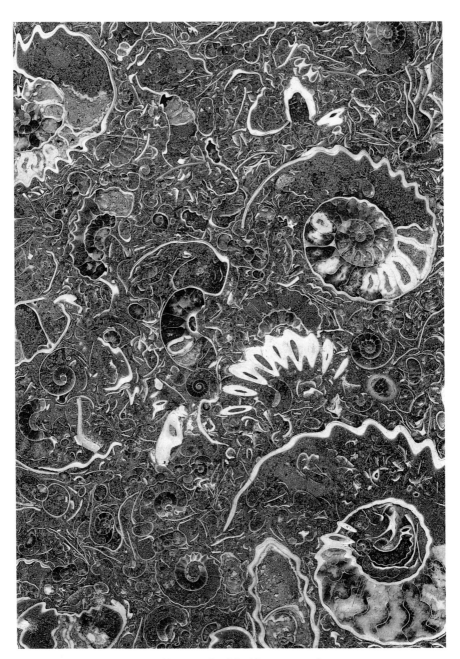

plate 21: Ammonite marble cut and polished for ornamental use.

Ammonites vary greatly in the length of the body chamber and the way the shell is coiled. There are some ammonites with short body chambers, about a third or a half of a whorl, but most have body chambers much longer than this. The Jurassic ammonite *Dactylioceras* is typical, with the body chamber occupying the entire final whorl of the shell. In such ammonites the aperture would have been oriented upwards. Such an ammonite would not have been able to see the sea floor, much less gather food from it. It is also unclear how such an animal could have swum, and so these ammonites have been interpreted as being plankton drifters. Working out how these ammonites would have been oriented depends on how big and how heavy the soft body of the animal was, something for which there is little evidence. For example, a heteromorph ammonite with an open hook-shaped shell such as *Hamites* can be restored with a body occupying the complete hook, in which case the aperture points upwards. But suppose that the body was much smaller, and like a snail it could move along the living chamber. With the body at the front of the living chamber, the shell would point forwards and only slightly away from the sea floor. If the ammonite was disturbed then it could pull itself deep into its shell for protection, and the aperture would tip upwards and away from the sea.

JET PROPULSION, HALF A BILLION YEARS BEFORE FRANK WHITTLE

Jet propulsion came quite late in the development of aircraft. Hot-air balloons similar to those still in use today were invented during the eighteenth century, and solid fuel rockets even earlier. Gliders sturdy enough to carry a human were in production by the close of the nineteenth century, and by the first decade of the twentieth century the addition of propellers driven by a lightweight internal combustion engine allowed the Wright brothers to make the first powered flight. It was not until the 1940s that Frank Whittle designed and built the first jet aeroplanes for the RAF towards the end of the Second World War, so it is perhaps surprising that the cephalopods have used exactly the same principles to get around quickly and efficiently for 500 million years.

A jet is in fact a very simple machine. It operates by taking fluid in at one end and compressing it. This pressurised fluid is then released through a nozzle, and the reaction against the rapid escape of this pressurised fluid drives the jet forward. In aeroplanes the fluid is of course air, while cephalopods use water.

In most aquatic molluscs the continual movement of cilia on the gills produces a gentle current between the mantle cavity and the outside, which refreshes the water inside the mantle cavity. This is important, since the gills extract the oxygen the animal needs from this water, while the kidneys and gut empty out the waste products of metabolism and digestion. For some molluscs, this gentle rate of mantle cavity circulation is insufficient. Mussels and oysters live in silty environments and frequently get clogged-up with the fine particles they draw in along with the plankton they eat. Periodically, they clap their two shells together, rather like castanets, rapidly flushing out the water in the mantle cavity along with mud or silt. Scallops have taken this clapping motion a step further, and use it in a much more regular and controlled manner, allowing them to swim. They can move fast enough to escape from their main predators, starfish. In the most advanced species, water is sucked in at the front and sides at low pressure, and then sent out in two high pressure jets from the back. They also have smooth, streamlined shells and two rows of tiny but sensitive eyes; these scallops are alert and surprisingly capable swimmers.

Cephalopods also push water through the mantle cavity using muscular contractions, but their system is more sophisticated than that employed by scallops. It needs to be: while scallops spend most of the time resting on the sea floor unless disturbed, cephalopods must move constantly to find food and avoid predators. In Chapter 1 we described how nautiluses use muscles to squeeze out water from the mantle cavity through the hyponome, to produce a powerful jet of water. The mechanism used by squids, cuttlefish and octopuses differs slightly from this. Instead of squeezing the water out of the mantle cavity, the mantle itself is able to contract in a very forceful manner. The water escapes through a nozzle, which in these cephalopods is a single flexible tube called a 'siphon'. When the mantle relaxes it expands, and water is sucked back in to refresh the mantle cavity.

Whether or not ammonites had jet propulsion, of either the nautilus or squid variety, is a matter of hot debate among palaeontologists. On the one hand they did have buoyant shells, like nautiluses, and retractor muscle scars are present on the inside of the shell indicating that they were able to pull the body into the shell. Also, as jet propulsion occurs in living examples of both primitive cephalopods (the nautiluses) and advanced cephalopods (the squids, and even the largely bottom dwelling octopuses), it is reasonable to suppose that jet propulsion was a characteristic of all extinct cephalopods, including the ammonites. However, the hyponomic sinus is not present in most ammonites, and in some species the apertural constriction or rostrum would have obstructed the hyponome, so perhaps ammonites did not have a hyponome. The spines and ribs on many ammonites would have greatly increased drag, as would the open coiling of some ammonites, particularly the heteromorphs. Finally, the body chambers of ammonites are totally unsuitable for producing an efficient and powerful jet of water. The mantle cavities of both nautiluses and squids are relatively short and broad, so that each ventilation cycle is brief but strong. In contrast the body chambers of ammonites are long and narrow, and in the case of many species of heteromorph, highly convoluted. These are the worst shapes for producing a good jet of water; the head would need to be pulled in a long way to flush out a useful amount of water and the bends in the chambers would waste energy instead of turning it into thrust.

STREAMLINING

Although living cephalopods have a powerful source of thrust in jet propulsion, the key to swimming efficiently is minimising drag. Perhaps the biggest obstacle to assuming that ammonites were adept swimmers is that so many of them seem to have had shells that produced lots of drag.

Drag is the name given to the forces which inhibit the movement of an object through a fluid. There are three main forces: friction, adhesion, and suction. Friction is the resistance of water against the forward motion of the object. As the animal moves through the water, the molecules of water rub and jostle against its surface, catching on the small imperfections on the animal's skin, slowing it down. Water is a 'sticky' or viscous fluid and adheres to any animal trying to swim through it; this adhesion is the second source of drag. Finally, as the animal swims forwards the space it occupied is instantly filled with water, being drawn in by suction. This suction is the negative pressure region where the animal was before it moved forwards – the empty space – and though it sucks in water, it also sucks back the animal. To keep moving, the animal must resist this force.

A swimming animal must therefore overcome all these forces if it is to keep moving. For very small animals these forces are insurmountably large relative to their own power. For example, small insects which accidentally land on the surface of a pool of water find it so 'sticky' they cannot take off again. For bigger animals, including cephalopods, the forces of drag are relatively weak, but they do add up. To have the greatest acceleration or to swim at a fixed speed using the minimum of energy, an animal needs a body shape that creates the minimum of drag; such shapes are said to be streamlined.

Friction and adhesion are dependent on the surface area of an object, the larger the surface area the more water molecules it disturbs and the more molecules can stick to it. Smooth objects have a lower surface area than rough or spiny ones, and so swimming animals tend to have smooth bodies without projections of any sort. (Certain very small projections reduce drag, by trapping a layer of water around the object to form a sort of lubricant. Shark skin has tiny tooth-like spines all over it, which seem to work in this way.)

Overall shape is important, not just texture, and rounded edges rather than square or flat ones reduce drag by reducing the turbulence produced as the object moves along. Less turbulence means that the suction force behind the animal is smaller.

Theoretically, the best shape for moving in water is one that is rounded in front and tapering at the back. Unsurprisingly, many fast moving aquatic animals have this shape, such as salmon, dolphins and penguins. Squid also have this shape, though compared with fish they are 'back to front', with the rounded tail-end leading and the head and arms trailing. When swimming at top speed, squids wrap their fins around their bodies and pull the arms into a tight cone, further improving their streamlining. Cuttlefish and octopuses are not as streamlined as squids, and they do not swim as quickly or for such long distances. Nautiluses are poorly streamlined compared with squids, but they are very slow swimmers and so streamlining is probably not important. At very low speeds the savings on energy gained by being streamlined are small and may not be worth it if they compromise other aspects of design. This is why the Japanese express 'bullet trains' are streamlined and the subway trains on the London Underground are not: at 250 km/hr^{-1} (155 mph) the savings are real and valuable; at 40 km/hr^{-1} (25 mph) they are irrelevant. On the other hand, if efficiency is important, then any savings, however marginal, are worthwhile, as can be seen in the smooth and rounded designs of most motor-cars produced today.

Historically, ammonites have been described as being good or bad swimmers on the basis of streamlining alone, irrespective of any other aspect of shell shape such as the length of the body chamber. For example, ammonites with globose shells covered in spines or ribs were considered to be unable to swim quickly, and so must have lived on the sea floor. Other ammonites with smooth, disc-like shells were supposed to be well streamlined and so must have been fast and efficient swimmers. This is now believed to be a gross oversimplification, and experiments using model ammonites in water tanks show that on the whole ammonites were about as well streamlined as the nautilus, or worse. None was anything like as well streamlined for swimming as the squids.

DEFENCE

In all molluscs with external shells, whatever else the shell does it must also provide a final line of defence against predators. At its very simplest a shell should be a barrier through which a predator cannot attack, and so it is not surprising that many molluscs have thick shells. The downside to having a thick shell is that it is heavy and takes longer to build than a thinner shell. A compromise reached by some molluscs, particularly marine snails, is to only thicken the shell in certain places while leaving the rest of the shell relatively thin. Although the thin part of the shell can be broken easily, the predator will hopefully be stopped when it reaches one of the thick barriers, called varices. Another solution is to make the shell too difficult for the predator to handle, thereby putting off the predator which will go and find an easier target. Spines and ribs are commonly found on snails and clams, as well as almost all ammonites, although surprisingly very few fossil nautiluses, and none of the living species. In some cases the spines may simply be a 'hedgehog' defence – by covering the entire surface and being long and sharp they make the animal inside an unpalatable mouthful. More often the spines are arranged in distinct rows along the shell and not all over, and so probably do not work in exactly that way. Instead they may make the shell seem bigger than it is, especially to predators that need to take the entire shell into their mouth or between their claws. If these sorts of predators cannot handle the shell properly, they cannot open it.

Many molluscs also use their shell to avoid detection by predators. Disruptive patterns break up the outline of the shell making the animal difficult to spot, particularly against complex backgrounds like coral or seaweed. Camouflage patterns also hide the outline of the animal by blending in with the surroundings. This is usually done by adopting the same colours and textures as the background. The nautilus seems to combine disruptive and camouflage patterns on one shell. Its red-brown stripes resemble those of the tiger, and break up its outline, but they are not uniformly distributed. Instead, they are boldest on the top surface and absent from the underside, the resulting pattern providing a form of camouflage known as countershading. From above, the nautilus is dark and so difficult to spot against the dark water below; while looking from underneath, the creamy white underside matches the light filtering down from above. The coleoids have internal shells, and they are well known for their colour-changing abilities. Small pockets of coloured pigments inside the skin can be expanded and contracted. The combined effect of thousands of these special cells is rather like the pixels on a computer screen, and almost any pattern or shade can be adopted.

Unfortunately, fossils very rarely preserve colours or patterns, and so we know next to nothing about what ammonite shells looked like in this regard. A few Tertiary nautiluses preserve traces of colour banding very similar to that of living nautiluses, as described in Chapter 1, suggesting that countershading might have been relatively common among externally shelled cephalopods. On the other hand, there are a few ammonite fossils which show bold markings, such as zigzags and spirals, and so these species at least might have relied on some kind of disruptive coloration.

Chapter Four

ASPECTS OF AMMONITE BIOLOGY

WE know very little about the food of ammonites compared with what we know about the feeding habits of living cephalopods. Many shallow water species of cephalopod have been observed in the wild or in aquaria, including squids, cuttlefish, octopuses and nautiluses. The feeding habits of the open ocean and deep sea species have been deduced from the stomach contents of some caught by net or line. This gives us a fairly firm basis from which we can interpret the evidence from the fossil record.

NAUTILUS

The nautilus is the logical animal to start with since it is the living cephalopod which most obviously resembles the ammonites. Like many of us, it is very fond of crustaceans, particularly large prawns and lobsters. This is borne out by the stomach contents of specimens taken from the wild, and also by the relish with which nautiluses feed on these animals in captivity. Nautiluses are opportunist feeders, meaning that they will take what they can find. Exactly how they forage is not completely understood since our observations of wild nautiluses are limited, but the sense of touch and smell seem to be most important. As it swims along, the

left: The Lower Cretaceous ammonite *Australiceras* sp. from Australia showing damage to the shell, producing a v-shaped track near the outer edge. The persistence of the track implies that the mantle was also damaged. Size, 8.5 cm 3.4 in) width.

nautilus sweeps its tentacles through the water and on the sea bed. Obviously this system has its limitations – any fish or shrimp which feels a nautilus tentacle can try to make a quick escape into its burrow – and the nautilus is probably not a very effective hunter. A peculiarity of the nautilus compared with squids and octopuses is that it eats the shells as well as the flesh, and will also eat the cast-off moults of large crustaceans even though they might seem rather unappetising. Possibly these shells are a valuable source of calcium which it needs to build its shell. Fossil nautiloids have been sectioned and found to contain pieces of crustacean, showing that nautiluses have been feeding on these animals for a very long time.

Cuttlefish

Cuttlefish have been kept in aquaria for many years, and their feeding habits are well known. Usually they settle on to the fine sand at the bottom of the tank and use their uncanny colour changing abilities to blend in perfectly, sometimes adding a sprinkle of sand over their backs to complete the disguise. If a small shrimp or fish is put into the tank, the cuttlefish will stay quite still until the animal moves into range, then, almost explosively, its two long arms shoot out and the little animal is captured. The whole process takes a fraction of a second – the arms shoot out in one hundredth of a second, and the prey is pulled back within one eightieth of a second. Facilitating this remarkable hunting method is particularly fine eyesight. Cuttlefish have binocular vision that allows them to see their prey and judge the distance and direction of their target before making an attack. The nautilus, with its poor, monocular vision could never do such a thing.

Octopuses

Octopuses also hunt on the sea bed but they tend not to wait patiently like the cuttlefish. They are active hunters and investigate every possible nook and cranny among the rocky habitats they favour in their quest for food. Like the nautilus and cuttlefish they are very fond of crustaceans. Despite their soft and clumsy appearance, octopuses are quite adept at catching and overpowering species as large as the edible crab, and sometimes even the formidable lobster falls prey. Octopuses will also eat large gastropods and clams, and even fish if they can catch one. A few species have been documented making brief excursions overland between rock-pools in search of prey, taking any prey they encounter en route – including rats! Octopuses feed in a distinctive way using their beak and a venomous saliva. With crustaceans and molluscs they use the radula to rasp a hole through the shell into which they inject the venom. The prey is paralysed almost at once, and the saliva can begin the process of digestion, loosening the muscles from the shell. After a while the octopus sucks the meat cleanly out of the shell, discarding the almost intact shell. In a few species this venom is powerful enough to be painful to humans, and one species, the blue-ring octopus, is deadly.

SQUIDS

Squids differ from all the other cephalopods as they feed not on the sea floor but in mid-water. Fish is a favoured prey item of many species, but swimming crustaceans such as krill are also taken. In fact squids are voracious and will attack almost anything that moves, including other squids. They rely on their great speed to approach their target quickly, and like the cuttlefish they have extendable arms which they use for snatching their prey quickly. Many species have sharp hooks on these arms, rather like cat's claws, for grabbing hold of fish which tend to be too slippery for simple suckers.

Not all squids are so active. One family, the Cranchidae, are passive drifters inhabiting relatively deep water. Instead of swimming they float, buoyed up by ammonium-rich tissues. Their bodies have a flabby, jelly-like appearance and a distinctly unpleasant smell. They are hugely abundant and form a major part of the diet of the deep-diving bottlenose and sperm whales, although to humans they are unpalatable. When oceanographers began mapping the sea floor they were surprised to see sonar echoes well above the sea floor, and that this 'deep scattering layer' as it was called, moved upwards at dusk and back down by dawn. The deep scattering layer is now known to be dense schools of fish, crustaceans and squids. During the daytime they inhabit deep water (between about 200 to 800 m (650–2620 ft)) away from many of the major predatory fish, seabirds and dolphins. When night falls these huge schools of animals swim upwards, into the top 200 m (650 ft) of the sea, where they can feed on the tiny planktonic plants and animals that are there. By daybreak the squids, crustaceans and fish have begun the return trip back to the safety of deeper water. This up and down cycle is called a diurnal vertical migration.

Ammonites

So much for the living cephalopods then, but what about ammonites? In Chapter 3 we discussed the swimming abilities of ammonites, which seem to have been limited at best. Also, while some contemporary cephalopods, such as belemnites, have been found with hooks suggesting that they had grasping arms like those of squids, no ammonite has ever been found with similar hooks. If ammonites could not pursue or catch fish, how did they feed, and what did they eat? In rare cases, ammonites have been found with what seem to be their last meals inside the living chamber. These remains include ostracods (tiny crustaceans around 1–2 mm in length or around 0.1 in), benthic foraminifera (single-celled organisms) and pieces of crinoid (filter-feeding echinoderms distantly related to starfish and sea-urchins). Broken up ammonite shells have also been found inside the living chamber, suggesting that at least some ammonites ate others of their kind. The problem is that we cannot be absolutely sure that what are interpreted as gut contents in ammonite fossils are actually that; the remains of other animals could easily have been incorporated as sediment filled the empty ammonite shell. At face value, at least, it would seem that while ammonites ate many different things they did not share quite the same tastes as modern cephalopods – the remains of fish, large crabs and lobsters are notably absent. In fact the list of animals they fed on can be divided into two groups: tiny benthic animals (ostracods and foraminifera) and larger but still slow moving animals (echinoderms and other ammonites).

The heteromorph ammonites are a special case. Virtually nothing is known for certain about their diet, any more than about their mode of life, but the shape of their shells do indicate that many could not have been powerful swimmers. Heteromorphs, such as the straight-shelled *Baculites* are shaped in such a way that the head would have hung straight down, and so if they could swim at all it would have been vertically and not horizontally. The obvious analogue among living cephalopods would be the cranchid squids with their diurnal vertical migrations. Did these ammonites move from deep water during the day into shallow water at night to feed? Heteromorph ammonites with helical shells, such as *Turrilites*, would also have hung head downwards, but even a gentle jet of water would have been enough to make them spin. Imagine a Cretaceous evening scene a few miles offshore at dusk. At a depth of 100 m (330 ft) or so, vast shoals of helical ammonites, each one 30 cm (12 in) or so in height, are slowly corkscrewing their way upwards. Each one is trailing its long tentacles, sweeping through the water and reeling in any tasty morsel it finds. Such a feeding method would have been very efficient, allowing them to search the surrounding water thoroughly for every scrap of food. For tens of millions of years these elegant, pirouetting predators must have been a very distinctive and beautiful part of the marine realm.

SEXUAL DIMORPHISM

During the nineteenth century, geologists recorded where they had collected fossils according to the standards of the time, often no more precisely than the 'upper', 'middle' or 'lower' part of a formation. Indeed, a large proportion of specimens in the older museums are labelled in this way. The problem with this is that the Upper Chalk, for example, was deposited over a period of millions of years. The twentieth century saw a progressive refinement of the stratigraphical divisions into smaller and more clearly defined intervals, some as brief as 20,000 years. Fossils collected from these narrow intervals produce assemblages of specimens which can be assumed to have lived more or less contemporaneously. Such 'bed-by-bed' collecting, as it is called, has proved especially interesting with regard to ammonites, showing that often two forms with very similar growth and shape characteristics occurred together, the two forms differing primarily in size alone.

TWO FORMS

At first, the two different forms were thought to be separate, though closely related, species. However, closer inspection of each form revealed a very curious thing: until they reached maturity, they were identical. The smaller variety, known as the microconch, looks very like the early whorls of the larger macroconch. It appears that the microconch and macroconch were growing in the same way, but the microconch stopped halfway through. Besides being bigger, the macroconch usually has longer body chambers, two-thirds of a whorl or more, whereas the body chambers of the microconch are only about half a whorl (see plates 13–14). A further distinction between the microconch and the macroconch is the shape of the living chamber, particularly the aperture. The microconch often possesses lappets, lateral extensions of the shell on either side of the aperture. Well developed lappets are typical of the late Jurassic ammonite *Kosmoceras*, the microconch having lappets equal in length to the diameter of the shell, which the macroconch lacks. The ornamentation on the later whorls of the macroconch is often weaker than it is on the earlier whorls, and on the final whorl may be absent altogether. In the late Cretaceous ammonite

Mortoniceras, the macroconch possesses a long, hook-like rostrum. Among the heteromorphs, the microconch of the straight-shelled heteromorph *Lechites* has a tapered aperture with fluted edges, while the aperture of the macroconch is curved over into a hood-like structure. So what were these two distinct 'varieties' of otherwise similar ammonites?

As mentioned in Chapter 1, the male argonaut is much smaller than the female, and a similar arrangement can be seen in other invertebrates, such as spiders. Although there is no conclusive evidence, many palaeontologists now believe that the two varieties of ammonite – microconch and macroconch – were in fact the male and female of a single species. The question is which was which?

WHICH IS WHICH?

If we look at living cephalopods, except for the nautilus, it is the females that are larger. Coleoid reproduction is very curious, and when these animals were first studied they caused much confusion. All coleoids have eight arms (plus two tentacles in the case of squids and cuttlefish), but in the males one of these arms is modified. This arm is called the hectocotylus and has a groove running along the inside. To fertilise a female, the male places a packet of sperm, called a spermatophore, inside this modified arm and then pushes the arm into the mantle cavity of the female. The arm then breaks off the male's body. He may remain close to the female to make sure no other male tries to mate with his chosen female, but otherwise his part of the mating process is over, and he dies soon after. The female breaks open the spermatophore when she is ready to fertilise and lay her eggs. During the nineteenth century, when naturalists first dissected cephalopods, once in a while they would find a female with a hectocotylus inside. They thought it was some sort of parasitic worm, which had taken on the appearance of a cephalopod arm as a disguise; they called these worms *Hectocotylus* ('arm with a hundred suckers'). Now we know that

AMMONITES

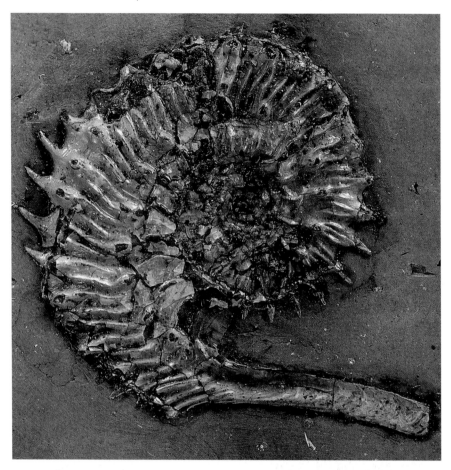

above: A microconch of the Middle Jurassic ammonite *Kosmoceras phaeinum*.

these are not worms at all, but the name has stuck. Since males only have one hectocotylus and one spermatophore, they can fertilise a female only once, so there is no need to defend a harem of females or defend a territory. On the other hand, the female's part in reproduction persists long after fertilisation. She produces relatively large eggs for a mollusc, and each egg is endowed with a large quantity of yolk. Female squids place the eggs inside tough cases which protect them from predators, while octopuses place the eggs inside a cave and stay with them until they hatch. But in all cases, the female dies soon after. This is an energetically expensive way of doing things, but does mean that when they hatch, the babies are perfectly formed little predators well able to look after themselves. In contrast, molluscs like clams and limpets cast large numbers of tiny eggs into the plankton. This means they have access to the microscopic animals and

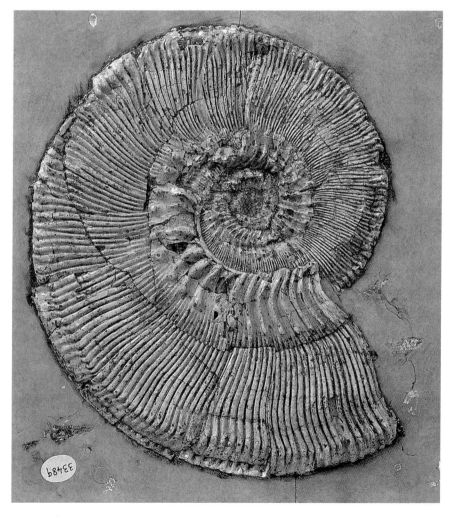

above: A macroconch of the Middle Jurassic ammonite *Kosmoceras phaeinum.*

plants there for food instead of the yolk sac – but they also run the risk of being eaten by the countless larger animals that feed on the plankton.

The nautilus is, as is so often the case, very different. Both sexes have dozens of thin arms, but the male has one large, spoon-like arm called a spadix with which he can place the spermatophore inside the female. Unlike the hectocotylus this arm does not break off, but can be used each time he mates. Female nautiluses produce only a very few eggs at a time and as far as we know they do not offer any brood care at all. Male and female nautiluses therefore have a much less demanding time when breeding, and can do so many times throughout their lives. They are also similar in size and shape, though the males are often slightly broader to accommodate the spadix. If the microconch and macroconch are truly different sexes of one species, then the stronger parallel is obviously with the coleoids rather than the nautiluses: the microconch was the male, and the macroconch the female.

But why?

Why were the sexes so strongly dimorphic? In many dimorphic species, the male develops secondary sexual attributes to display its overall health and vitality, in other words, its quality as a potential mate. The peacock's train, the tail-fins of male guppies, the tusks of narwhals are all examples of such attributes. Beyond impressing females, they serve little purpose, indeed, they can be a liability, for example the bright colours on the tail-fins of guppies are known to make them easier targets for predators than the drab females. The females prefer the more brightly coloured males, possibly because to have survived long enough to reach sexual maturity, such males must be particularly alert or fast swimmers. Did the lappets of the microconch work in the same way, showing off to the females which males were the best? Many palaeontologists believe so, and the idea that the lappets were 'claspers' which held on to the female while they were mating, is now largely discounted.

If the male was elaborately ornamented to impress the female, then why was the female, the macroconch, so much larger than the male? If the macroconch is the female, why did they need to be bigger than the male and why were their living chambers proportionally larger? What did they use all this extra body space inside the shell for? As described in Chapter 1, most octopuses brood their eggs inside caves, and one kind of octopus, the argonaut, goes even further and builds a shell-like egg case especially for this purpose. In this way the female can protect the clutch of eggs from predators, remove any that are unfertilised, and keep the eggs clean and free from parasites or fungi. Female ammonites may have done a similar sort of thing by brooding the eggs inside their spacious shells. If they did, then such a strategy is unusual for cephalopods. Nautiluses do not brood their eggs, and generally nor do cuttlefish or squids.

AMMONITE REPRODUCTION

In contrast with many other molluscs, which are hermaphrodite, all living cephalopods are either male or female. As described above, it seems likely that this was the case for the ammonites too, with the microconch and macroconch probably being the male and female respectively. However, there are important differences in the reproductive biology of nautiluses and coleoids, and one significant gap in our understanding of ammonites is which group they more closely resembled. Nautiluses grow quite slowly and it takes several years, perhaps as many as ten, for them to become fully grown. Sexual maturity might occur sooner, after about five years or so, but virtually nothing is known about how nautiluses reproduce in the wild. Nautiluses may live for 15 to 20 years, and will mate and lay eggs many times throughout their lives. In captivity nautiluses lay only a few eggs at a time, and these eggs can take many months to hatch, but the eggs are large and the hatchlings relatively large, perfectly formed miniature versions of their parents. Coleoids are very different. They can grow extremely quickly, many species reaching adulthood within a year. Once fully grown, they mate only once, and then die. The males may die immediately after spawning, or else guard the female until she has laid her eggs. The females may die after laying the eggs, as is the case with squids and cuttlefish, or, as described earlier, they may protect the eggs until they hatch, as octopuses do. This is one of the great mysteries of coleoid biology – why such large, advanced and behaviourally sophisticated animals should be 'annuals'. Did ammonites 'live fast and die young' like coleoids, or did the grow slowly and reproduce many times like nautiluses?

Clues may come from the way the shell developed. Ammonite shell growth can be divided into three distinct phases. The first is that of the protoconch, the single barrel-shaped chamber which is believed to have been formed inside the egg, and so corresponds to the embryonic phase in the life cycle of the ammonite. The protoconch of an ammonite is much more like the protoconch of a coleoid

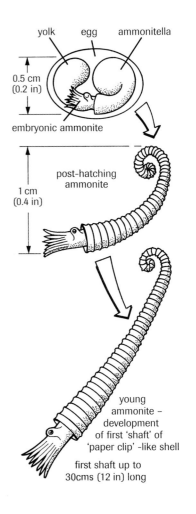

above: Ammonite shell growth phases.

than that of a nautilus. The next stage of growth was the first whorl, which ran from the protoconch to a distinctive change in shell morphology called the primary constriction. The first whorl is relatively plain compared to the later part of the shell, though it may have a simple form of ornament such as pimples. There are no spines or ribs. The protoconch and the first whorl together comprise the ammonitella. It is believed that the ammonitella was planktonic, rather than free swimming or benthic. After the primary constriction, which appears to correspond to a hiatus in shell secretion, the shell morphology changes, acquiring the mother-of-pearl, or nacreous, inner walls and septa typical of adults. This is the third and final stage of growth, during which time the ammonite left the plankton and took up its adult mode of life. The external ornamentation changes markedly, the simple pimples of the protoconch and first whorl being replaced with fine growth lines, and later ribs, spines and keels. Subsequent growth is essentially uniform from this point onwards, excepting the sexually dimorphic features described earlier.

At hatching, nautiluses are quite large, with a shell diameter of around 2 cm (1 in). Although there is a growth hiatus, called the nepionic constriction, similar to the primary constriction seen in ammonites, there is no drastic change in shell morphology or mode of life. Ammonite hatchlings – the ammonitellas – are much smaller, and so it appears that ammonites produced large numbers of small eggs, much as modern squids and octopuses do, rather than a few big eggs as is the case with nautiluses. Here at least the similarities in ammonite life cycles are strongest with the coleoids. But did they follow the coleoid pattern of having a fast growth rate, a short lifespan, and breeding only once? The simple answer is we do not know.

AMMONITE OLD AGE, PATHOLOGY AND PREDATORS
Mature ammonites are easy to identify. Sexually dimorphic ones develop the lappets or collars at maturity, as well as any distinctive ornamentation. Even species with little or no dimorphism show signs of a slow down in growth after maturity. The septa become 'bunched' as the new chambers become smaller. But being able to tell when an ammonite reached maturity is not the same as being able to say how old it is. So how long did ammonites live for?

above: The Upper Jurassic ammonite *Peltoceras dorsetensis*, showing simple ribbing on the phragmacone and spinous ribbing on the body chamber. This change in ornamentalism is typical of mature ammonites.

Lifespan

Without any living ammonites to observe, our estimates of the length of ammonite lifespans are somewhat vague. The living species of nautilus grows slowly, adding a chamber every month or two, and taking several years to reach sexual maturity, after which they grow much more slowly. Although they do not stop growing, the later chambers are noticeably 'bunched', that is, the septa are much closer together than they are in the rest of the shell. Ammonite shells are broadly similar to those of nautiluses, and so ammonites may have grown at a similar rate to nautiluses. If this was so, then simply counting the number of chambers would give a reasonable estimate of their lifespan. The problem is that we do not know how similar the growth of modern species of nautilus is to ammonites, or even other, now extinct species of nautiluses. Modern nautiluses live in deep, cold water. Growth rates in invertebrates are directly related to ambient temperature and pressure; low temperatures and high pressures result in low growth rates. Ammonites appear to have preferred shallow, warm water, and so might be expected to have grown much faster than modern nautiluses.

An indirect growth rate indicator was described by the influential German palaeontologist Schindewolf in 1934, who published a report of a Jurassic ammonite encrusted with tubeworms. This paper was important because it tied together something we can observe today, the growth of tube worms, with something unknown, the growth rate of ammonites. Schindewolf noticed that on his specimen of the ammonite *Schlotheimia* an encrusting serpulid worm had grown its tube around the ammonite's shell in such a way that the two animals must have been alive at the same time. Schindewolf knew the growth rate of these worms, and so could estimate the growth rate of the ammonite. He thought that each whorl of the ammonite shell took somewhere in the range of four months to three years to grow. A typical ammonite can have a dozen whorls, and so we are looking at a lifespan of anything from four to 36 years. However, while Jurassic species of tube worms might have grown at the same rate as their relatives alive today, we cannot be certain.

Ammonite shells contain isotopes of certain elements such as oxygen and carbon. Isotopes are varieties of the same element which differ in atomic weight, and can be detected using certain analytical techniques. The odd thing about the isotopes of oxygen in particular is that they occur in different proportions depending on the ambient temperature. By carefully analysing the composition of the shell it is possible to detect the relative abundance of each isotope of oxygen at the moment that piece of shell was being laid down. In this way, each part of the shell acts as a thermometer set to the exact moment it was made! Palaeontologists have found that if pieces of shell are taken from points all along the shell, different oxygen isotope proportions are found. In fact, it seems that as the ammonite grew it was sometimes in warm water, sometimes in cooler water, and then back again to warmer water, with this cycle happening many times. The most obvious conclusion is that these are the seasons – the periods of growth recording cold phases during the winter, and warmer phases in summer. By adding up these cycles, we can tell how many years the ammonite lived for. *Borrisiakoceras*, for example, a small Cretaceous ammonite, has been estimated to have lived for about 12 years. These various age estimates greatly exceed those of the coleoids but are comparable to the living nautilus. It would seem, therefore, that ammonites grew more slowly than modern coleoids, and so the short life span and single breeding event typical of coleoids seems unlikely to have been the case with ammonites.

PATHOLOGY AND PREDATORS

A related and very interesting aspect of ammonite palaeontology is the study of old, sick and damaged ammonites. After reaching sexual maturity, ammonites, like nautiluses continued to grow albeit at a reduced rate.

left: An enlarged view of the scar damage on the Lower Cretaceous ammonite *Australiceras* sp. shown on p. 88.

Compressed septa are not uncommon, but in addition reduced or deviant ornamentation can also be seen. Such specimens are called 'gerontic specimens'. Ammonites also show evidence of illness and damage from predators. Parasitic infections may have caused the blisters visible on the shells of some ammonites such as the heteromorph ammonite *Baculites*. Some ammonites survived attacks from predators but were damaged sufficiently to require substantial shell repair. In other cases the ammonite was not so fortunate, and the attack killed the animal and destroyed the shell. The remains of the shell can give clues to the nature of the attack, in particular, what sort of animal the predator was. Tooth marks on some specimens of *Placenticeras* match the jaws of marine reptiles, indicating that these animals were predators on ammonites.

Chapter Five

AMMONITE TAXONOMY AND CLASSIFICATION

THE Ammonoidea are one of the three subclasses or orders within the Cephalopoda, and are characterised by having chambered shells with complex septal walls, a ventral siphuncle, and a radula with nine teeth per row. The group is further subdivided into suborders, which are defined on the basis of features such as suture line complexity, the shape and position of the siphuncle and septal necks, and overall shell shape and ornamentation.

Many of these groups give rise to others over time, and frequently the more primitive groups are completely replaced by their more advanced descendants. The orders are the Anarcestina, Clymeniina, Goniatitina, Prolecanitina, Ceratitina, Phylloceratina, Lytoceratina, Ancyloceratina, and Ammonitina.

There is some inconsistency in the spelling of these orders. Some authors prefer the use of the ending -ida rather than -ina, for example Ceratitida instead of Ceratitina. In this book we use the ending -ina. A further complication is in the use of the word 'ammonite'. Strictly, the term is used for a single order, the Ammonitina or 'true ammonites', in contrast to the term 'ammonoids'

left: The rich variation of ammonites.

which refers to all the members of the Ammonoidea. In practice this limited use is seldom adhered to by any but specialists, and in this book we have chosen to use the broader, if less accurate, meaning of ammonite as any member of the order Ammonoidea.

Four suborders are exclusively Palaeozoic: the Anarcestina, Clymeniina, Goniatitina and Prolecanitina, and are the most primitive of all the ammonites. They show a little more complexity in the shape of the suture line than do contemporary nautiloids or coleoids, and typically have smooth or finely ribbed shells without spines or keels. Other characteristics are variable and resemble the other cephalopod groups. In the Anarcestina, Goniatitina and Prolecanitina the siphuncle is ventral, as it is in coleoids, while the Clymeniina have a dorsal siphuncle like some nautiloids. Similarly, the septal necks of the Anarcestina and Clymeniina are retrochoanitic, like nautiloids and most coleoids, while those of the Goniatitina and Prolecanitina are prochoanitic as is the case with the later ammonites. As a consequence of both the greater age of these four suborders, and their relatively primitive morphology compared with the Mesozoic ammonites, the Anarcestina, Clymeniina, Goniatitina and Prolecanitina are placed in an informal grouping known as the palaeoammonoids, or 'ancient ammonites'.

Only a single suborder, the Ceratitina, makes up the next group, the mesoammonoids or 'middle ammonites'. These are predominantly Triassic in age, although a few are from the Permian. Compared with the palaeoammonoids these ammonites have more complex suture lines, and a consistently ventral siphuncle with prochoanitic septal necks. Perhaps more markedly than any of the other groups, the Ceratitina are a grade group, or stage, in the evolution of the higher ammonites, bridging the gap between the Palaeozoic Goniatitina and the Mesozoic Ammonitina, Ancyloceratina, Lytoceratina and Phylloceratina.

The remaining Mesozoic ammonites make up the neoammonoids, or 'modern ammonites'; namely the orders Ammonitina, Ancyloceratina, Phylloceratina, Lytoceratina. All these ammonites have complex suture lines, and like the mesoammonoids the siphuncle is always ventral and the septal necks prochoanitic. In addition, the shells of these ammonites are frequently ornamented with ribs, constrictions and spines. The Ancyloceratina are commonly known as the 'heteromorph ammonites' on account of their diverse, and frequently bizarre, shell shapes which are quite unlike the regular spirals of most other ammonites.

ANARCESTINA

The Anarcestina developed from the Bactritina during the early Devonian, and retain some features common to that group, such as a simple septa and retrochoanitic septal neck. Unlike the Bactritina, however, all Anarcestina have coiled spiral shells and a ventral siphuncle, two features typical of ammonites in general. For this reason, the Anarcestina are considered the first true ammonites. Compared with later ammonites the Anarcestina are not particularly diverse, and the various genera can be quite difficult to tell apart. Like the Bactritina, the Anarcestina have simple suture lines with only a few lobes, a large external lobe and a smaller internal lobe. In some species there are only these two lobes, but in others there is an additional lobe on each flank, between the external and internal lobes. This is the lateral lobe, and is usually larger in size than the internal lobe.

Although all the Anarcestina had died out by the end of the Devonian, they gave rise to three other ammonite groups. One was very short lived, the Clymeniina, which also became extinct at the end of the Devonian. The second group was the Prolecanitina which was rather more diverse than the Clymeniina, and much more successful, lasting into the Triassic. The third and largest group was the Goniatitina and their descendants, which included all the Jurassic and Cretaceous ammonites.

above: The Upper Palaeozoic *Agoniatites* sp.

left: The suture line of *Anarcestes lateseptatus* from the Middle Devonian of Germany.

Most of the Anarcestina have little ornamentation, usually limited to fine, sinuous ribs, and they do not have spines. Coiling is typically convolute, with a dorso-ventrally compressed whorl section which expands very slowly, resulting in a globose, compact shell. The aperture lacks lappets or rostra, but does have a hyponomic sinus very similar to that found on the modern nautilus, indicating that, like the nautilus, these ammonites had a moveable hyponome through which water could be jetted for propulsion and orientation. The ammonitella is quite large, about 1 mm (0.1 in) in diameter, and ornamented with very fine transverse ribs or lirae.

The ecology of the Anarcestina, like most of the Palaeozoic ammonites, has not been studied as intensively as the Mesozoic ammonites. The Anarcestina are relatively abundant in oxygenated, shallow water sediments perhaps indicating a preference for the coastal waters of the continental shelves. The fact that these ammonites have a well developed hyponomic sinus is probably a good indicator that they were active swimmers, but the relatively globose shell of many species would seem to preclude great speed.

CLYMENIINA

This small group of ammonites is limited to rocks of Upper Devonian age, and is distinguished from all other ammonites by the position of the siphuncle in the shell. If a clymeniid ammonite is sectioned, the siphuncle

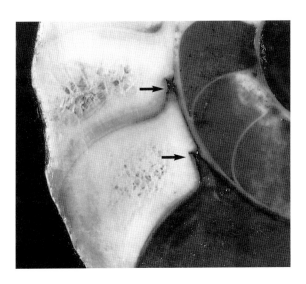

left: Sectioned *Clymenia* showing the dorsal siphuncle (the backwards pointing channels near the top edge of the chambers indicated by arrows).

above: *Gonioclymenia laevigata* from the Upper Devonian of Germany.

can be clearly seen. In the very first chamber, just like most other Palaeozoic ammonites, the siphuncle is in a ventral position close to the outside of the spiral. But for the rest of the shell, the siphuncle is along the dorsal, inner surface.

Although a relatively small group with only a few species, the Clymeniina are surprisingly diverse. For example *Solicylmenia* has a bizarre triangular form, while *Epiwocklumeria* is very involute, almost globose with deep constrictions. *Clymenia* is much more typical, being a relatively evolute regular spiral with a circular or dorso-ventrally compressed whorl section. The surface is smooth, the only ornamentation being fine growth lines. Like the Anarcestina, the Clymeniina have a hyponomic sinus, indicating they had a flexible hyponome for jet-propelled swimming. The suture lines of these ammonites are relatively simple, and resemble those of the goniatites described below. Clymeniina are only weakly sexually dimorphic, if at all.

Fossils of Clymeniina can be found in a variety of sediments, including limestones, clays and black shales, and are common in the nodular limestones of Morocco where they are often collected and sold as curios. Given the general requirement of abundant oxygen to supply the high metabolism of cephalopods, it is unlikely they were able to spend much time in anaerobic water, so where they do occur in black shales it is most likely that the shells sank down after death; the animal itself lived high up in the water column where there would have been plenty of oxygen.

Like the Anarcestina, the Clymeniina failed to survive the Devonian mass extinctions which wiped out many other marine invertebrate groups. They do not seem to have given rise to any other ammonite groups; all the later ammonites have ventrally positioned siphuncles. Rather than being a basal stock from which later groups emerged, the Clymeniina seem to have been something of a short-lived experiment.

GONIATITINA–THE GONIATITES

The Goniatitina, commonly called goniatites, are the most important of the palaeoammonoids, having a wide variety of morphologies and numerous species. Some species were widely distributed and lasted for only a relatively brief period of time, making them useful for biostratigraphy. Characteristic of goniatites is a simple suture line with triangular external, lateral and internal lobes and rounded saddles in between them. This particular sort of suture line, the so-called goniatitic suture line, is also exhibited by many of the Anarcestina, Clymeniina and Prolecanitina and is typical of the most primitive ammonites. However, in the Goniatitina and Prolecanitina particularly, there are many species which have complex suture lines, and resemble the ceratites much more strongly.

above: *Goniatites crenistria* from the Lower Carboniferous of the UK, showing rounded saddles and pointed lobes.

left: The suture line of *Anarcestes* sp.

Goniatites are variable in size and shape, but most are rather small, relatively compact and have involute, smooth or weakly ribbed shells. A number of species have longitudinal or spiral ribs which run along the shell. The aperture lacks keels or lappets though there may be a modest hyponomic sinus. The ammonitella is relatively small, with an initial chamber less than 0.7 mm (0.03 in) in diameter, with a short, curved prosiphon. The siphuncle sometimes starts from a central position and gradually moves to a ventral one, but is usually close to the ventral surface throughout growth. The ammonitella is ornamented with longitudinal ribs (lirae) like those of the adult shell.

Fossils of goniatites are common in sediments deposited under both aerobic and anaerobic conditions, and can be found in many different types of rock. Some species are most abundant in limestones, and seemed to prefer shallow water, while others, found in black shales, lived in the deeper water but away from the anoxic sea floor. Having smooth shells and a hyponomic sinus, goniatites could have been quite mobile swimmers. The more complex suture lines of later species indicates that in some cases goniatites were able to live in relatively deep water.

PROLECANITINA

Compared with the other palaeoammonoids, the Prolecanitina are characterised by a generally flattened shell with rather complex suture lines and numerous lobes and saddles. The Prolecanitina appeared in the late Devonian but died out by the end of the Triassic, being replaced by their descendants, the Ceratitina.

The early Prolecanitina have evolute or involute shells, with more or less circular whorl sections, and are laterally compressed to a lesser degree than the later species, which all have involute, disc-like shells. The surface of the shell is generally smooth or only weakly ornamented with ribs; there are no constrictions or lappets. The suture line is much more complex than in the other palaeoammonoids. The saddles are rounded while the lobes are almost always divided into small finger-like projections. The suture lines of the more primitive species are like those goniatites, but the more advanced species show more complexity. In all of the Prolecanitina there are external, lateral and umbilical lobes, but in the advanced species as growth progresses an additional lobe appears between the lateral and internal lobe beneath the umbilical region; this is the umbilical lobe. This umbilical lobe may itself further subdivide, so that the suture line on later chambers can be very complex, with numerous small lobes between the lateral and umbilical lobe (see p. 44). Such a suture line is very characteristic of the advanced palaeoammonoids; it can be seen in some goniatites and is also typical of the ceratites as described below. More complex suture lines make the shell more robust and resistant to both the compressive forces of deep water, and damage caused by predatory attacks. The trend towards increasing suture complexity in the palaeoammonoids may be related to either the expansion of ammonites into deep water habitats or in response to more dangerous predators, or indeed both.

above: The Lower Carboniferous *Prolecanites compressus.*

In many of the later species the ventral face is not curved or flattened like many of the other palaeoammonoids but tapered into a sharp edge. This gives the shell an overall form rather like a discus. A similar shape is also seen in some of the Jurassic and Cretaceous ammonites, and is believed by some palaeontologists to be an adaptation to fast, or at least efficient, swimming.

CERATITINA-THE CERATITES

Compared with the palaeoammonoids, the Ceratitina, generally known as ceratites, show a greater diversity of shell shape and external ornamentation, and a much more complex suture line. Many are laterally compressed, and weakly ornamented, resembling the Prolecanitina described above, but there are numerous species which have strong ribs or tubercles, while others have a rounded or quadrate whorl section, giving the shell a much more robust appearance. There are also some heteromorphic forms, resembling the Ancyloceratina in having open or uncoiled shells. Some of these are not coiled in one plane, but are instead helical, like a snail's shell.

The ammonitellas of ceratites are small, with an initial chamber less than 0.7 mm (0.03 in) in diameter, and are ornamented with small round tubercles. The prosiphon is short and curved, and the first part of the siphuncle either starts from a central position and gradually moves to a ventral one in later chambers, or is close to the ventral surface all along. The most characteristic feature of the Ceratitina is the ceratitic suture line. There are generally four lobes, the external and internal lobes, and between them a large lateral lobe and smaller umbilical lobe. In some species, as the ceratite grew larger additional lobes were added by division of the umbilical lobe, as with the Prolecanitina. Unlike the Prolecanitina, which started off with simple sutures with three lobes and only developed the umbilical lobe as they grew larger, all Ceratitina began with at least

AMMONITES

above: *Ceratites nodosus* from the Middle Triassic, showing rounded saddles and divided lobes.

left: The suture line of the Lower Triassic *Paraceratites elegans*.

four lobes. The lobes are usually quite complex, while saddles remain rounded and simple.

Although there are Permian ceratites, which appear to have been derived from the Prolecanitina, ceratites are more typical of the Triassic. Some are found in black shales, implying they lived in the upper levels of relatively deep water, above anoxic bottom water conditions; but mostly they are found in limestone, indicating a preference for the warm, shallow continental seas fringing the land masses of the time. In the Lower Triassic marine limestones of China, distinct assemblages of ceratite species have been identified, indicating that not all ceratites liked the same conditions and that different species of ceratites became specialised for different habitats. The robust, strongly ornamented species with less complex suture lines are found in shallow water conditions (less than 100 m or 330 ft depth), often associated with burrowing bivalves, suggesting soft bottoms of sand and silt. More streamlined species with rounded, laterally compressed shells and weak ornamentation are found in deeper water conditions, as are the heteromorphic ceratites, along with a diverse array of bivalves, gastropods, brachiopods and crustaceans.

The most streamlined species, with oxyconic shells and little or no ornament are found in the deepest water environments. They also have the suture lines of the greatest complexity, and seem to have been able to tolerate depths of up to 200 m (650 ft). This diversity is significant in showing the remarkable recovery of the ammonites following the mass extinctions at the end of the Permian. As described in Chapter 6, the Permian saw the end of many of the Palaeozoic animal groups, such as all the trilobites, many brachiopods and corals, as well as many of the ammonites. Only a very few of the Prolecanitina and Ceratitina survived, and soon after, the Prolecanitina became extinct. The ceratites, in contrast, gradually diversified as the conditions became more amenable, and as the shallow water habitats recovered, so these new ammonites soon became important predators within these ecosystems.

PHYLLOCERATINA

Appearing in the late Triassic, the Phylloceratina is one of the most uniform and conservative groups of ammonites. The suborder includes only a single superfamily within the Phylloceratina, the Phylloceratacea, within which all the known species are remarkably similar. These ammonites are rather large, and the shell tends to be robust, more or less involute and with rounded whorl sections. They are also streamlined in appearance, being either smooth or ornamented with very fine ribs, and all are laterally compressed. Many species have periodic constrictions on both the phragmocone and the living chamber but they do not have keels. The aperture of some species have lateral and ventral extensions, but these are gently curved and rather short, and not at all like the strongly developed lappets and rostra of many Ammonitina.

The suture line is finely divided compared with the Ceratitina, with characteristically spiky lobes but rounded, lifelike saddle elements. The primitive species have a suture line that is quadrilobate, being divided up into external, lateral, umbilical and internal lobes. In Jurassic and Cretaceous species there are usually more than four lobes. These additional lobes, often referred to as 'adventitious' lobes, were formed by divisions of the umbilical lobe as ontogeny (growth) progressed. Like all the Mesozoic ammonites, the siphuncle is ventral and the septal collars are prochoanitic. All Phylloceratina have small ammonitellas with small initial chambers, and are ornamented with tubercles as with the Ammonitina and Lytoceratina. The prosiphon is either central or marginal.

The complex suture lines indicate that these ammonites had robust shells and may have been better adapted to life in deeper water than the

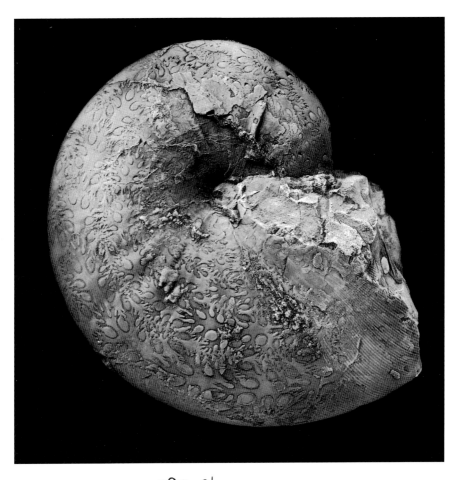

above: The involute shell of *Phylloceras heterophyllum* from the Lower Jurassic.

left: A Phylloceratid suture line with lobes and saddles diminishing equally.

Goniatitina or Ceratitina. Having streamlined, smooth and involute shells it is thought that the Phylloceratina were fairly good swimmers, though they lacked keels and none were as laterally compressed as, for example, *Oxynonticeras* or *Placenticeras*, which are both true ammonites, members of the Ammonitina (see p. 128).

As the only order of ammonites to occur in both the Triassic and the Jurassic, the Phylloceratina are assumed to be the link between the Triassic Ceratitina and the Jurassic and Cretaceous Ammonitina, Ancyloceratina and Lytoceratina, although the exact relationships are not clear. Their most obvious similarities are between the Phylloceratina and the Lytoceratina, which both have weakly ornamented shells with periodic constrictions, and both groups maintain a similar morphology throughout their long stratigraphical ranges.

The Phylloceratina are widespread and known from all around the world from the late Triassic until the very end of the Cretaceous. They are very common in the Jurassic deep water Ammonitico-Rosso limestones common around the Mediterranean, where they can be by far the commonest ammonites. On the other hand, they tend to be very rare or absent from the shallow water habitats favoured by other ammonite groups.

Being rather uniform in shape, fragments of these ammonites can be very difficult to put a name to. Many a palaeontologist has found a collection of such ammonites, and in despair called all the specimens *Phylloceras* species. Furthermore, even the best specimens show that many species changed very little over long periods of time, compared with the other, more rapidly evolving ammonite groups.

LYTOCERATINA

A major suborder of ammonites, the Lytoceratina were one of the most persistent and widespread of all the Mesozoic ammonite groups. Believed to have been derived from the Phylloceratina, from which they differ in being more evolute, the Lytoceratina are known throughout the Jurassic and up until the very end of the Cretaceous. On the whole the Lytoceratina are relatively conservative despite the long period of time during which they occurred.

The Lytoceratina are divided into two superfamilies, the Lytocerataceae and the Tetragonitaceae. The former are most abundant in the Jurassic and tend to be evolute serpenticones with simple annular ribs periodically thickened into flared collars, and with numerous constrictions all along the shell. A few of the Lytocerataceae, such as *Pictetia*, have shells which are so loosely coiled that the whorls are not in contact with each other. The Tetragonitaceae are more common in the Cretaceous, and all have regular, involute shells, but are otherwise similar to the Lytocerataceae. All Lytoceratina share a common ammonitella morphology. The ammonitella has a short, curved prosiphon, and the outside of the ammonitella is ornamented with small, rounded tubercles. The size of the ammonitella is variable, with an initial chamber from 0.3 to over 1.0 mm (around 0.1 in) in diameter, smaller than the Palaeozoic ammonites but similar to, or slightly larger than, the other Mesozoic ammonites. As with the other Mesozoic ammonites, the siphuncle runs along the ventral surface of the shell and is bounded by prochoanitic septal collars. Sexual dimorphism is less common, and not as marked as among the Ammonitina, and where present is limited to small differences in size. The key diagnostic feature of the Lytoceratina is a complex suture line with a distinctive division of the lobes and saddles. The most primitive species such as *Trachyphyllites*, is quinquelobate. Two umbilical lobes (named U_1 and U_2) are formed by the split across the umbilical seam of the initially singular umbilical lobe. In later Lytoceratina, such as *Lytoceras*, there may be three or more of these umbilical lobes formed by further subdivisions.

The circular whorl sections and evolute shells of typical Lytoceratina suggest that these poorly streamlined ammonites would have been rather indifferent swimmers. The open umbilicus would be a major source of drag, leading to poor acceleration and making rapid lunges at prey, or quick escapes from predators, difficult. The more involute species of Tetragonitaceae might have been better swimmers.

AMMONITES

These ammonites are found in a wide range of sedimentary habitats, like the Phylloceratina. The deep water limestones of Jurassic age around the western Mediterranean are particularly rich in ammonites of the

above: Evolute Lower Jurassic shell of *Lytoceras fimbriatum*.

left: Lower Jurassic Lytoceratid suture line.

Phylloceratina and Lytoceratina groups. The Lytoceratina seem to have been specially adapted to life in deeper water than most other ammonites. They have closely packed septa which are quite thick, indicating that the shell was able to resist the greater hydrostatic pressure of life in deep water. On the other hand, even the late Cretaceous Lytoceratina have weakly ribbed or smooth shells, which is odd when compared with the general trend among the Mesozoic ammonites towards stronger ornamentation. This may indicate that the diversification of predators like crabs and bony fish during the Jurassic and Cretaceous was not a major influence on the evolution of the Lytoceratina.

One particular characteristic of the Lytoceratina, especially Lytocerataceae such as *Pictetia*, is the tendency towards uncoiling, a trend which reaches its greatest expression among the Ancyloceratina, an order of ammonites which evolved from the Lytoceratina during the late Jurassic and went on to become one of the most diverse and successful of all the ammonites.

ANCYLOCERATINA–THE HETEROMORPHS

The Ancyloceratina included some of the most widespread, and some of the most common, of all ammonites and they are an important group for biostratigraphy. They remained widespread throughout the late Cretaceous, in contrast to the Ammonitina, which were in a marked decline. The Ancyloceratina are commonly known as the 'heteromorph ammonites' on account of their uncoiled shells (see plates 15–17). Although rarely used, the term 'homomorph ammonites' exists for ammonites with regular spiral shells, such as the Ammonitina.

There are five main groups of Ancyloceratina: the Ancylocerataceae, Deshayesitaceae, Douvilleicerataceae, Scaphitaceae and Turrilitaceae. The most primitive of the Ancyloceratina, the Ancylocerataceae appeared during the late Jurassic and typically have shells which are basically spiral

above: *Nipponites mirabilis* from the Upper Cretaceous of Japan, showing tangled coiling.

but with a protracted hook-shaped living chamber hanging beneath. The Deshayesitaceae and Douvilleiceratace are two groups of heteromorphs which revert to regular spiral coiling and tend to have evolute, strongly ribbed, and often quite spiny, shells. Both groups were most common and widespread during the Cretaceous, where they provide some of the most biostratigraphically useful ammonite species.

The Ancyloceratace, Deshayesitaceae and Douvilleiceratace became extinct during the Cretaceous, when the advanced heteromorphs, the Turrilitaceae and Scaphitaceae, diversified. The Turrilitaceae are the older of these two groups, and radiated explosively in the Cretaceous into an extraordinary array of shapes. The Scaphitaceae, like the Deshayesitaceae and Douvilleicerataceae, are characterised as a group by the reversal of 'heteromorph-ness' and have closed, often quite involute, spiral shells. Unlike the Deshayesitaceae and Douvilleicerataceae, most Scaphitaceae have a slightly uncoiled living chamber resulting in a distinctive scoop-like shallow living chamber. While it is now generally accepted that the Scaphitaceae were derived from the Turrilitaceae during the Cretaceous, the relationship between the Ancylocerataceae and Turrilitaceae is obscure. It is possible that rather than the Ancylocerataceae evolving steadily through the early Cretaceous and culminating in the first of the Turrilitaceae, the Turrilitaceae and the bulk of the Ancylocerataceae share a common ancestor.

However they may differ in coiling mode, all Ancyloceratina share a quadrilobate suture line similar to those seen in the Lytoceratina. The suture line is divided into external, lateral, umbilical and internal lobes, at least during the juvenile stages of growth. In many cases the suture lines of later chambers show subdivisions of some lobes, particularly the umbilical lobe, into what are called 'pseudolobes'.

above: *Douvilleiceras mammilatum* from the Lower Cretaceous, showing ribs with many spine bases.

above: Fragments of *Hamites* sp. from the Lower Cretaceous.

AMMONITINA–THE TRUE AMMONITES

The Ammonitina are the largest ammonite group, containing many more species than any of the other groups. They appeared very early in the Jurassic, but whether the first Ammonitina were most closely related to the Ceratitina, Lytoceratina or Phylloceratina is not clear (see plates 18–20). The Ammonitina were especially common and diverse in shallow and moderate depth environments, and if their shell morphology is any reflection, they occupied a variety of ecological niches.

Ammonitina have complex suture lines. The earliest chambers have five lobes but as growth progressed some of the lobes, commonly the umbilical lobes, became further subdivided. The lobes and saddles are almost always deeply incised but are never leaf-shaped like the Phylloceratina, and are more like the same features of the Lytoceratina. Current estimates suggest that there are 15 superfamilies within the Ammonitina, within which there are approximately 1000 known genera.

AMMONITE TAXONOMY AND CLASSIFICATION

above: *Parkinsonia dorsetensis* from the Middle Jurassic.

left: The suture line of the Upper Cretaceous *Mortoniceras inflatum* showing typical development of the lateral lobe.

right: The first Jurassic ammonite to appear in the UK, *Psiloceras planorbis*, from near the base of the Lower Jurassic.

Although many of the Ammonitina are quite disparate in size and shape, they do all share a number of features in common. Coiling is generally that of a regular spiral, though a few are heteromorphic. Dimorphism is very common, many species being divided into the smaller microconch and larger macroconch, presumed to be the male and female of the species respectively. The microconch often has secondary sexual attributes such as lappets, while the macroconch lacks these but may have other distinctive features such as apertural collars or constrictions. The ammonitella is usually small, the initial chamber being less than 0.7 mm (0.03 in) in diameter, with a long and straight prosiphon followed by an approximately central initial siphuncle and ornamented with round tubercles.

Among the Jurassic forms evolute, smooth shells are common, while the Cretaceous ammonites tend to be much more involute and strongly ornamented with ribs and spines. This has been explained as a response to an increase in the diversity and efficiency of predators such as fish and crustaceans. Though ribs and spines may have made these ammonites better able to survive the attacks of their enemies, the increased drag they produced would have made swimming more difficult. A few of the Ammonitina, such as *Oxynonticeras* and *Amaltheus*, had smooth oxyconic shells, and may have been adapted for swimming in the open sea, but most of the Ammonitina preferred to live on or near the sea floor.

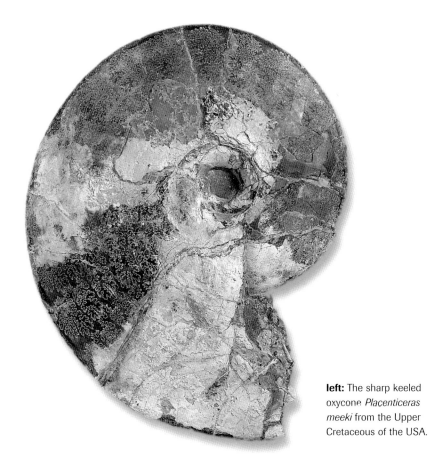

left: The sharp keeled oxycone *Placenticeras meeki* from the Upper Cretaceous of the USA.

Two particular features of the Ammonitina are the short lifespan, and the tendency of many species or families to have a limited geographic distribution. Short lived species of ammonites, in some cases estimated to have lasted only a few hundred thousand years, which are also widely distributed, can be used to correlate different rock formations if the same

above: *Kosmoceras duncani* with the suture line clearly visible.

above: Lateral and ventral view of the sharp keeled involute *Oxynoticeras lymense*.

species occurs in both. Also, groups of abundant species of ammonites which only occur in a limited area can be used to define 'faunal provinces'. These are most useful where continental drift has distributed the rocks bearing these ammonites widely, since it demonstrates that at one time they must have been part of a distinct and continuous marine basin. The Hoplitidae, a family of Cretaceous ammonites including the familiar spiky ammonite *Hoplites*, defines a region covering much of northern and western Europe which lay under an isolated sea during the mid-Cretaceous.

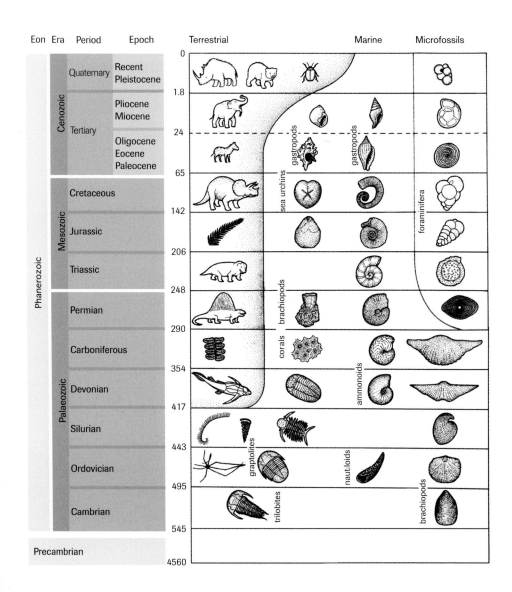

Chapter Six

THE EXTINCTION OF THE AMMONITES

THE entire history of ammonites can be summarised as a series of radiations and extinctions. Indeed, soon after their appearance in the Devonian, a mass extinction event wiped out many of the most primitive groups, including the Anarcestina and Clymeniina. This event is known as the Frasnian–Fammenian boundary event (after the two geological stages that bound it) and the ammonites were not the only victims, many brachiopods, nautiloids and trilobites also became extinct. The few ammonites that did survive, such as the Goniatitina and Prolecanitina, found themselves in a world devoid of competition and radiated explosively into a range of new and ever more successful types.

PERIODIC EXTINCTIONS THROUGHOUT THE PALAEOZOIC AND MESOZOIC

left: The main divisions of geological time (figures are in million years). The geological timescale was originally established from the relative ages of the rocks, partly based on fossils. Some of the important fossils for dating are shown.

This Palaeozoic paradise was not to last. An even greater mass extinction at the end of the Permian killed off virtually all the ammonites. This extinction event also signalled the end for many typically Palaeozoic invertebrate groups such as the trilobites, while relegating many others to the minor roles they occupy today, as with the brachiopods and nautiluses. After the Permian mass extinction, the animal groups that became dominant included many that remain important today, including the reptiles, bivalve molluscs, and crustaceans. A few ammonite groups did survive, notably the Phylloceratina and the Ceratitina, and these diversified throughout the Triassic.

By the end of the Triassic the Earth was passing through yet another period of mass extinction. This time the Ceratitina failed to survive, but the Phylloceratina made it through, and the succeeding geological periods (the Jurassic and Cretaceous) saw a diversity and abundance of ammonites that had never been seen before. The Lytoceratina, Ancyloceratina and Ammonitina all have their roots in those few species of Phylloceratina that made it through the Triassic–Jurassic boundary extinctions. There were further extinctions in the Jurassic and Cretaceous, but they were not so severe as the preceding ones, and all four ammonite groups lasted until the end of the Mesozoic.

During the 1980s, two American scientists – Raup and Sepkoski – observed that from the Permian extinctions onwards, each extinction event happened at an interval of about 26 million years. They hypothesised that this was because the extinctions were caused by some periodic astronomical event. What this event might be is open to speculation. One idea is that the Sun has a companion star (nicknamed Nemesis, after the Greek goddess of righteous vengeance) that every 26 million years passes close enough to the Solar System to fling comets from beyond the orbit of Pluto crashing in on the planets, including Earth. Such impacts would be catastrophic, the celestial equivalent of a nuclear war. Another suggestion is that the Sun is not as constant as it has been through human history, with a cycle of lethal periods of variability between long intervals of benign constancy. Yet another idea concerns the orbit of the Sun, and thus the Solar System itself, through the Galaxy. Might the Solar System periodically pass through regions of the Galaxy where conditions are more hazardous, exposing the Earth to cosmic rays or other deadly forms of energy?

Fascinating as all this might seem, other scientists are far from convinced. Closer inspection of the fossil record has shown that the various extinctions do not perfectly match the 26 million year cycle, and that before the Permian none of the extinctions match it. Moreover, in seeking evidence for extinctions some scientists exaggerated the scale of the events, while other events, including the Permian extinction, clearly took many millions of years to work through. One of the most contentious examples of all was the end-Cretaceous mass extinction at which the ammonites became extinct.

LATE CRETACEOUS DECLINE AND EXTINCTION

The end of the Cretaceous saw the complete extinction of many animal groups, and is probably the best known of all mass extinctions. The dinosaurs were the most dramatic victims. After virtual dominance of the land for over 100 million years, they vanished, geologically speaking, over night. The ammonites, too, took their final bow around this time, as did the belemnites. This period of mass extinction as been named the Cretaceous–Tertiary, or K/T, boundary extinction event (K is the geological symbol for the Cretaceous – from the German 'kreide' meaning chalk – and T is used for the Tertiary). While geologists argue about the celestial causes of some of the mass extinctions, there is excellent evidence for an astronomical event on Earth at the end of the Cretaceous period: the Chicxulub crater on the Yucatán Peninsula in Mexico. The Chicxulub crater is now more or less submerged beneath younger rocks, but it can be revealed using special devices sensitive to magnetic fields – the geological equivalent of an x-ray machine. The crater is vast, perhaps as much as 200 km (125 miles) in diameter, and it is believed to have been caused by a meteoritic impact. In comparison, the famous Meteor Crater in Arizona is only 1.2 km (0.75 miles) across.

The Earth is continuously being bombarded with dust and rock. Most of this material is small and burns up in the atmosphere. These are meteors, or shooting stars, of which on any given night a careful observer is likely to see a few every hour. As these pieces of dust fly into the atmosphere they are incinerated, producing a brief flash. Meteorites are much larger rocks which are not burned up by their descent, and such rocks can be found all over the Earth's surface. Where they hit the ground they produce craters. Several large meteorites are known, weighing up to 60 tonnes (55 tons) and measuring up to about 10 m (32.5 ft) across. In comparison the meteorite that produced the Chicxulub crater is estimated to have been over 10 km (6.2 miles) in diameter and must have weighed thousands of tonnes. Something this big smashing into the Earth would have had a devastating effect. Some of the debris from the impact would have been hot enough to start fires, and layers of ash found in many places around the world at about this time do seem to indicate that some parts of the Earth were indeed ravaged by forest fires. On top of this, it seems to have crashed into the worst possible place. Beneath the crater are layers of sulphur rich rocks, and the combination of the heat from the meteorite and the shock of the impact could have vaporised this sulphur to produce clouds of acid rain.

Some scientists agree that there is a good argument for tying the K/T extinctions to the impact of this meteorite. The problem is that for the ammonites, and many other animal groups, the K/T boundary did not mark a sudden extinction of diverse groups. The ammonites were a group in steep decline long before the K/T boundary. Over 20 families of ammonites are known from the mid-Cretaceous, but by the late Cretaceous there were only half as many, and only a few of these persisted to the end of the late Cretaceous. A similar decline in numbers can be observed as well, with late Cretaceous ammonite fossils being quite scarce. Unless the ammonites knew 'their end was near' and decided to retire gracefully, it is very difficult to see how the meteoritic impact could have been the immediate cause of ammonite decline and extinction. Many other animal groups show a similar pattern. The great marine reptiles, the mosasaurs, ichthyosaurs and plesiosaurs, had either declined severely or become extinct long before the end of the Cretaceous, as had flying pterosaurs. One very distinctive group of molluscs known as the rudists, which were very diverse through much of the Cretaceous, had also become extinct well before the end of the late Cretaceous. Even the dinosaurs were a shadow of their former glory by this time, and the latest dinosaur bones known to science still pre-date the K/T mass extinction by tens of thousands of years. For these reasons many palaeontologists have turned to slower mechanisms as explanations for the K/T mass extinctions, such as long term climate change or large scale volcanic activity. While not as dramatic as a meteoritic impact, such phenomena would have acted over the millions of years needed to cause the gradual extinction of so many groups.

THE AMMONITES' SUCCESSORS-THE COLEOIDEA

Although most palaeontologists have traditionally made comparisons between the living nautiluses and the extinct ammonites, more recently the coleoids have become just as important. The coleoids include the familiar squids, cuttlefish and the octopuses, as well as a number of more obscure living and fossil cephalopods. Also numbered among the coleoids are the belemnites, extinct squid-like animals with abundant fossils in sediments of Jurassic and Cretaceous age, which like ammonites have become useful for biostratigraphy.

Modern cephalopods are diverse in external appearance, but are more or less similar to belemnites in basic anatomy, and so their close affinity is not questioned. Squids are perhaps the most obviously belemnite-like cephalopods. From the outside, the most apparent difference is that whereas belemnites had ten arms of similar size and shape, the arms of squids are differentiated into two sorts. There are eight short arms, which like belemnites might be covered in hooks, but may alternatively have suckers, or in some cases a mixture of both. In addition there are two very long arms, each ending in a club-shaped region bristling with spines or suckers. Usually these tentacles are hidden within the arms, and only emerge when feeding. They can be extended with incredible speed, and are used to snatch the small fish and shrimps that squids enjoy eating. Like belemnites, squids have internal shells, but these shells are much less well developed. There is no rostrum or phragmocone, and the entire shell is more like a long, thin feather; indeed, the squid shell is often called a 'pen', as the shell resembles the old-fashioned writing quills made from the feathers of large birds like geese. The shell is not calcified, but is instead made from a flexible organic material (partly calcified chitin).

Externally very similar to the squids, cuttlefish have much more complex internal shells. Whereas the shell of the squid is nothing more than a flexible 'backbone' keeping the body straight and stiff, the shell of the cuttlefish is more like that of a belemnite. There is a true phragmocone, though it is highly modified, and instead of a series of more or less cylindrical chambers like those of belemnites, the shell is divided into thousands of tiny compartments. This arrangement allows the animal to adjust its density quickly and efficiently. Both squids and cuttlefish have a streamlined shape much like a belemnite, although cuttlefish tend to be somewhat stocky. They also have strong, muscular fins, which they use to swim with when they are in no particular hurry, reserving jet propulsion for emergencies.

The octopus is probably the best known living cephalopod. It has a body shaped like a bag, and from the inside of the bag extend the eight arms that give it its name. There are no fins. Between the arms and the opening of the bag is the 'head' region, perched on top of which are two strikingly well-developed eyes. Although the entire animal can be quite large, octopuses are famous for being able to squeeze through the smallest openings. They can do this because they really only have two 'hard parts', the jaws and the brain-case. The jaws are much like those of other cephalopods, and the brain-case is small. The rest of the body is soft, and so long as it can squeeze the jaws and the brain-case through a hole, it can get its entire body through.

Why the ammonites became extinct at the end of the Cretaceous, while the nautiluses and coleoids survived, will probably never be fully understood. It remains a mystery not only because the ammonites survived other extinctions earlier in their history, but also because some groups, particularly the heteromorphs, showed little sign of being an obsolete or unadaptable group. Quite the reverse in fact, the heteromorphs seemed to be occupying new ecological niches that ammonites had hitherto not exploited. Whatever the reason for their extinction, the cephalopods never again held the pre-eminence in the world's oceans that the ammonites had. Though coleoids are diverse, and successful up to a point, they are overshadowed by the vertebrates in virtually every case, the bony fish in particular far exceed the coleoids in diversity and abundance. Compared with the long heyday of the ammonites, the coleoids enjoy only an echo of that success.

Collecting ammonites and ammonite collections

Being widely distributed and frequently abundant, it is not difficult to build up a good collection of ammonites. The pre-requisite to a successful collecting trip is planning. Ammonites are normally only found in Devonian to Cretaceous marine sediments. Geological maps are useful in finding likely exposures of these sediments. Amateur geological societies often run field trips to localities with particularly fossil-rich sediments, and inexperienced collectors will find these very rewarding. Ammonites can be extracted and cleaned in many different ways, depending on the sediments that the fossil was interred in, and so exactly how a fossil should be removed from the sediment will vary. Generally it is best not to use force, and not to try and clean the fossil too much in the field. It is much easier to remove unwanted bits of rock or clay at home where you can be more careful not to damage the fossil. Ensure that a note is made of where the fossil was collected from. Once at home the fossils can be cleaned and a labels written with all the collecting data recorded. Fresh fossils are often damp and should be allowed to dry off slowly. After they are dry, they should be quite durable and can be displayed, the main exception being fossils preserved in iron pyrites (fool's gold). If such a fossil was collected from a coastal exposure, soaking the fossil for a few days in water to remove salt helps slow down its decay, but otherwise pyritic fossils should be kept in a dry, airtight container. No fossil should ever be painted with varnish; not only does it obscure the surface detail but it also prevents the natural movement of moisture out of the fossil.

Further information

SCIENTIFIC LITERATURE

Aspects of ammonite biology, biogeography, and biostratigraphy. W. J. Kennedy & W. A. Cobban. Special Papers in Palaeontology, 17, 1976. [Although rather dry, the quality of the information included in this review of ammonite palaeontology is very high.]

Treatise on invertebrate palaeontology Part L. Mollusca 4. W. J. Arkell, W. M. Furnish, A. K. Miller, R. C. Moore, O. H. Schindewolf, P. C. Sylvester-Bradley & M. K. Howarth. Geological Society of America and the University of Kansas Press, 1957. [The series from which this book is a part is more commonly known simply as the 'Treatise'. It is the first reference book most professional palaeontologists turn to when they need information on the classification of fossil invertebrates. This particular volume includes some detailed summaries on the anatomy and ecology of ammonites at the time of publishing.]

Treatise on invertebrate palaeontology Part L. Mollusca 4 (Revised). C. W. Wright, J. H. Callomon & M. K. Howarth. Geological Society of America and the University of Kansas Press, 1996. [This is a recently revised section of the 1957 'Treatise' concerning the classification of Cretaceous ammonites, and now published as a separate volume.]

Books on ammonites and other cephalopods

Ammonites. Patrice Lebrun. Minéraux et Fossiles, Hors Série 4, December 1996. [Written in French, this is an excellent semi-popular review of the origins of the Ammonoidea and the anatomy of the ammonite shell and soft body parts, and includes many very useful diagrams and charts.]

Ammonites 2. Patrice Lebrun. Minéraux et Fossiles, Hors Série 6, December 1997. [The companion volume to *Ammonites*, and concentrates on the ecology, evolution and extinction of the Ammonoidea.]

The ammonites: Their life and their world. Ulrich Lehmann. Cambridge University Press, 1981. [The English language translation of Lehmann's classic book on ammonites. This is now out of print.]

Kingdom of the Octopus. Frank W. Lane. Jarrolds, 1957. [Probably the best-known account of the cephalopods. Although out of print for many years, it is well worth scouring the shelves of second-hand book stores for. Still a good read after over 40 years, the author displays a contagious enthusiasm for the cephalopods, and offers a wealth of interesting information and keen observations which will fascinate all readers.]

General topics

Archimedes & the Fulcrum. Paul Strathern. Arrow Books, 1998. [A short but delightful biography of the mathematician Archimedes, including simple explanations of his most important contributions to science.]

Cat's Paws and Catapults. Steven Vogel. Penguin Books, 1998. [A detailed but easy to read account of the functional morphology of animals and plants, particularly the engineering of systems such as the skeleton and locomotion devices. The attractive twist to this book is the comparison between natural 'living machines' and the devices built by engineers and architects.]

From the Beginning. Brian Rosen and Katie Edwards. Natural History Museum, London, 2000. [A general introduction to geology and the history of life on Earth.]

The Hidden Landscape. Richard Fortey. Jonathan Cape, 1993. [Richly illustrated and an easy read, this book is a good introduction to the geology and fossils of the British Isles.]

The Natural History of Shells. Geerat Vermeij. Princeton University Press, 1993. [A beautifully illustrated review of the way animals, primarily molluscs, make their shells and the way the design of the shell is adapted to its lifestyle.]

Websites

NB. Website addresses are subject to change.

http://is.dal.ca/~ceph/TCP/index.html
[Gives an overview of the Cephalopods with information on forthcoming conferences, photographs, and links to databases.]

http://perso.wanadoo.fr/jean-ours.filippi/anglais/
[General introduction to ammonites, with specific pages dedicated to the host's French Jurassic collections.]

http://www.humboldt.edu/~natmus/Exhibits/FossilTypes/Ammonites/WhatAmm.html
[Hosted by Humboldt State University, California, this site gives a general introduction to ammonites, their forms, suture patterns and abundance with beautiful photographic examples.]

Glossary

Ammonite   In general usage, this is the common name for all the Ammonoidea (see below). As used by palaeontologists, however, this name is used to describe only the most highly evolved members of the Ammonoidea, namely the Phylloceratina, Lytoceratina, Ammonitina, and Ancyloceratina. The more primitive Ammonoidea are instead referred to as goniatites or ceratites.

Ammonoidea   One of the three major groups of cephalopods, but extinct since the end of the Cretaceous period. Ammonoidea are in general characterised by having much more complex suture lines than either the Nautiloidea or Coleoidea.

Anaptychus   A single semicircular shaped structure associated with the aperture in some species of ammonite, thought to be made of an organic material called conchiolin. The plural is 'anaptychi'. The function of these structures is uncertain. The anaptychus may have been part of the jaw of the ammonite, or alternatively some sort of 'door' for sealing off the shell (see aptychus below).

Aperture   The opening at the front of the body chamber. This is the hole through which the head, arms and the hyponome (if present) emerged from. A peculiarity of many species of ammonite was the constriction or ornamentation of the aperture when they became fully grown. Sometimes within a single species such modifications differed between the two dimorphs. The aperture is sometimes called the 'mouth' of the shell, although it is obviously not the mouth of the actual ammonite animal.

Aptychus A structure made of calcite associated with the aperture of the shell. Commonly, the Mesozoic ammonites had a pair of these structures (the plural is 'aptychi'). In some fossils the aptychi fit snugly in the aperture and so look to have served as 'doors', perhaps sealing off the animal from the outside when it was threatened. On the other hand, like the anaptychus, the paired aptychi do resemble the jaws of living cephalopods, and may have served the same purpose.

Aragonite A form of crystalline calcium carbonate ($CaCO_3$) used by most cephalopods, including ammonites, as the basic material for the shell. Aragonite is also used by other molluscs including many gastropods, tusk-shells, and some bivalves. Aragonite is relatively unstable, that is it tends to decay over time, commonly to a more stable form of calcium carbonate known as calcite. Consequently, fossil ammonites rarely have this original shell material intact.

Bactritina Commonly, bactritids. An obscure group of straight-shelled nautiloid-like cephalopods that had their heyday in the Silurian and Devonian. Virtually nothing is known about their biology and ecology, but they seem to have been the ancestors of both the Ammonoidea and the Coleoidea.

Body chamber Also known as the living chamber. This was the chamber within which the soft body of the ammonite resided. It was the largest chamber, and always the final and outermost one of the entire whorl. It was connected to the buoyancy chambers via the siphuncle, and opened to the surrounding sea water at the aperture.

Cephalopoda The cephalopods. The third largest group of molluscs and the group within which the ammonites belong. Living cephalopods include a few species of pearly nautilus, dozens of species of cuttlefish, hundreds of species of squids and the octopus. There are even more species of cephalopods known only from the fossil record, including the ammonites and the belemnites.

Chamber   The ammonite shell is divided up into a series of chambers by dividing walls called septa. Most of the chambers were used to maintain neutral buoyancy, but the largest and final chamber was the body chamber.
Chitin   A tough but flexible organic material used by many organisms for parts of their skeletons. Insects, for example, have skeletons largely made up of chitin. Among cephalopods chitin is used extensively, for example the beaks and claws of living species are chitinous.
Coleoidea   One of the three major groups of cephalopods and the one with the most species alive today. Cuttlefish, squids and octopuses are all coleoids, as are the extinct belemnites.

Dimorphism   Some ammonite species show dimorphism, with two distinct forms being present, known as dimorphs. Commonly the main distinction is size, but they may also have different ornamentation on the shell or additional structures such as lappets. Sexual dimorphism is common among those animal species in which the males and females have radically different roles. Among ammonites many species show strong dimorphism (see macroconch and microconch below).
Dimorphs   Literally, 'the two forms' of a single ammonite species. Generally the dimorphs differ in size, one being smaller (the microconch), than the other (the macroconch). The smaller is assumed to have been the male and the other female, although this is pure speculation.
Dorsal   The back surface of an object. In the case of an ammonite the shell generally coils over the dorsal surface, though in heteromorphs this is difficult to see sometimes. Consequently, the dorsal surface of a single whorl of the shell is the surface towards the centre of the spiral.

Endogastric   Coiling over the dorsal surface of an animal. Nautilus and ammonite shells usually coil endogastrically.
Exogastric   Coiling under the ventral surface of an animal. The shells of coleoids such as cuttlefish commonly coil in an exogastric manner.

Haemocyanin   A respiratory compound containing copper found in the blood of many invertebrates, including cephalopods, and similar to the iron-based haemoglobin in human blood. Haemocyanin has a bluish colour. In cephalopods, the haemocyanin is not found in corpuscles but is dissolved in the blood plasma. Its main purpose is to bind with oxygen from the sea water as it passes through the gills, and then to yield that oxygen to tissues elsewhere in the body where required. Compared with haemoglobin, haemocyanin is less efficient and carries less oxygen. This may be one reason that cephalopods lack the stamina of similar sized fishes.

Heteromorph   Most ammonites have spiral shells which are basically the same shape throughout growth. Such ammonites are termed homomorphs (meaning 'the same-shaped shell'). A major group of ammonites appeared during the late Jurassic, the Ancyloceratina, which instead had uncoiled shells. These were the so-called heteromorph ammonites ('different-shaped shells'). Common shapes among the heteromorphs included helically coiled shells, long and straight ones like elephant's tusks, and even ones which looked like tightly folded paper clips.

Hyponome   A pair of overlapping, muscular flaps extending from the mantle which are held tightly together to produce a short and flexible tube. Only present among the living cephalopods, and used to direct the jet of water expelled from the mantle cavity. Often called the 'funnel'.

Lappets   Lateral extensions of the shell on either side of the aperture. These are generally found in the microconch only, and may have served as markers of sexual maturity and virility, similar to the peacock's tail.

Lobe   The posterior-pointing folds in the septum where it meets the inner surface of the shell. An important part of the suture line and used in classifying ammonites.

Macroconch   Within a dimorphic species, the larger of the two forms. By comparison with other strongly dimorphic cephalopods such as the paper nautiloid (Argonauta) it is thought that these were the females. Compared with the production of sperm it is the production of eggs or live young which requires the greatest energy reserves and largest body size, providing a strong selection pressure in favour of large females.

Mandibles   The jaws of an animal; in cephalopods often similar in shape to the bill of a parrot or predatory bird, and so commonly called the 'beak'. In ammonites the anaptychus and aptychi have been considered by some to be part of the mandibles.

Mantle   The outer layer of the body of a mollusc which secretes the shell.

Microconch   Within a dimorphic species, the smaller of the two forms. In many sexually dimorphic invertebrates the smaller of the two sexes is the male. Since production of sperm places few demands on body size and energy reserves, there is selection in favour of small but numerous males.

Nautiloidea   The earliest and most primitive of the three major groups of cephalopods, of which only a few species are alive today, such as the pearly nautiloid *Nautilus pompilius*.

Neutral buoyancy   The property of an object that neither sinks nor rises when placed in water. Many cephalopods have chambered shells which are partially emptied so reducing their overall weight and thereby making their density close to that of sea water. In this way they become neutrally buoyant, which reduces the energy needed to swim and gives them great control over poise and stability.

Phragmocone   The chambered part of the shell of a cephalopod, which by reducing overall density provides neutral buoyancy.

Plankton   The animals (zooplankton) and plants (phytoplankton) which drift in the open sea. Juvenile ammonites were probably planktonic and may have fed on other planktonic animals. A few ammonites may have been plankton throughout their entire lives, notably the heteromorphs.

Prosiphon   The beginning of the siphuncle. The prosiphon is located within the first chamber of the shell.

Protoconch   Literally the 'first shell'. This is the initial chamber of the phragmocone and in living cephalopods is produced before the animal hatches from the shell. The shape and size of the protoconch is important in classifying ammonites.

Radula   Row of rasping teeth on a tongue-like structure behind the mandibles of many molluscs including cephalopods. They are made of chitin.

Saddle   Anterior-pointing folds in the septum where it meets the inner surface of the shell.

Septa   The walls dividing the shell into chambers. The singular is 'septum'.

Siphon   A muscular nozzle present in most coleoid cephalopods used to direct the jet of water while swimming. Similar to the hyponome of the nautilus.

Siphuncle   A thin tubular extension of the back of living *Nautilus* to the origin of the shell. It regulates the buoyancy of the whole animal. It is present in fossil nautiloids and ammonites, and is believed to serve the same function.

Suture line   The pattern made by the lobes and saddles of a single septum around its entire edge. Generally the suture line is simple for the first few septae, and becomes more complex with successive ones. Groups of closely related ammonite species often have similar suture lines, and so the suture line is useful in classifying ammonites.

Umbilicus   The recessed centre of the shell of an ammonite or nautilus as viewed from the side.

Ventral   The underside of an object. The ventral surface of a single whorl of an ammonite shell is the outside edge of the spiral.

Index

Page numbers in *italic* refer to captions, those in **bold** to colour plates. Page numbers prefixed by (g) refer to an entry in the glossary.

abyssal plain 60-1
adhesion force 84-5
Aegocrioceras quadratus 38
age 101
    lifespan 103-4
Agoniatites sp. *110*
Allonautilus 23
Amaltheus 131
    A. gloriosus 40
ammonitellas 14, 100-1, 115, 117, 120, 123, 130
ammonites (g)148
    reconstruction *49*
Ammonitina (true ammonites) 109, 128-33
Ammonoidea 12-13, 18-20, (g)148
Anahoplites 38
anaptychus 45, 48, (g)148
Anarcestes 114
    A. lateseptatus 110
Anarcestina 108, 109-11
anatomy
    ammonites 48-51
    nautiluses 25-31
'ancient ammonites' *see* palaeoammonoids
Ancyloceratacae 125-6, 127
Ancyloceratina 109, 125-8
    see also heteromorphs
apertural constrictions *see* rostra
aperture 35, 37, 39, 64, (g)148
    dextral/sinistral **75**
aptychi 14, 45-7, 48, (g)149

aragonite 13, 27, 33, 36, 57, (g)149
    see also mother-of-pearl
Archimedes' Principle 56
architecture, ammonite-inspired 9
Architeuthis 48
Argonauta 21
    A. hians 22
argonauts (paper 'nautiluses') 21-2, 25
    sexual dimorphism 95
Aristotle 24
arms
    ammonites *35*, 48
    coleoids 95
    nautiluses 28, **66**
    squids 139-40
    see also hectocotylus
Aspidoceras 45
Asteroceras 53
    A. stellare 38
Aturia **71**
Aulacostephanus autissiodarensis **72-3**
Australiceras 89

Bactrites carinatus 19
Bactritina 18, (g)149
Baculites **76**, 93
beak *35*
    nautiluses 28
    see also mandibles
belemnites 12, 13, 21, 36, 92, 137, 139
Belemnoteuthis antiquus 20
bifurcate ribbing *39*
biostratigraphy 15, 113, 125, 126, *135*, 139
blood 51
    see also haemocyanin
body (living) chamber **79**, (g)149
    nautiluses 27, **65**

INDEX 155

Borrisiakoceras 104
Bostrychoceras polyplocuin 75
brachiopods 67
Brasilia bradfordensis 29
buoyancy
    centre of 18, 63, 64
    early molluscs 15
    neutral 14, 55, 58-9, (g)152
    principle of 56-7

cadicone shells 36, 37
calcite 17, 33, 45, **68**
Cambrian period nautiloids 17
camouflage
    spiral lines **71**
    tiger stripe 24, 50, **71**, 87
    see also coloration and patterning
Carboniferous period
    Lower Carboniferous ammonites *114, 116*
Cephalopoda 11-15, (g)149
    classification *12*
    history 15-23
Ceratites nodosus 118
Ceratitina (ceratites) 108, 117-20
chambers **68**, (g)150
    see also body chamber
chitin 36, 140, (g)150
classification *see* taxonomy and
    classification
Clymenia 111, 113
Clymeniina 108, 111-13
coiled nautiloids 18
coiling 36-9
    endogastric (g)150
    exogastric 25, 34, (g)150
    tangled *126*
    uncoiling 125
Coleoidea 13, 20-3, 139-41
coleoids 20
    colour-changing 87
    reproduction 95, 99-100
    see also cuttlefish; octopuses *and* squids
collecting fossils
    bed-by-bed 94
    collections 143
Collignoceras woolgari **70**
coloration and patterning
    ammonites 41, 87

nautiluses 27, 87
    see also camouflage
colour changing
    coleoids 87
    cuttlefish 90
conchiolin 148
countershading
    nautiluses 27, 87
counterweighting 17
Cranchidae 91
Cretaceous period ammonites 122, 126
    Lower Cretaceous *38, 41, 88, 127, 128*
    Upper Cretaceous **70, 74, 75, 76**, *126,
    129, 131*, 125
Cretaceous-Tertiary (K/T) boundary
    extinction event 137, 138
crinoids 67, 92
crushing, degrees of 34
cuttlefish 13, 50, 55, 85, 140
    buoyancy 20
    colour changing 90
    cuttlebone 61
    eyesight 90
    feeding 90
    reproduction 95
    water pressure resistance 61
cyrtocone ammonites *37, 38*

Dactylioceras 81
    D. commune 9, 38
damage scarring *89, 105*
defence against predation 40, **71**, 86-7, 131
Deshayesitaceae 125, 126
Devonian period ammonites 109
    Upper Devonian *111, 115*
dimorphism *see* sexual dimorphism
dimorphs (g)150
displacement, water 56, 57
disruptive coloration
    see coloration and patterning
diurnal vertical migrations 92, 93
dorsal surface (g)150
Douvilleiceras mammilatum 127
Douvilleiceratacae 125, 126
drag 84-5

echinoderms 93
egg cases 21, *22*

eggs
    ammonites 99, *100*
    nautiluses 36
embryonic ammonites *100*
embryonic shells 19, 36
endogastric coiling (g)150
Eoasianites 45
Eoderoceras 45
epifauna 31
Epiwocklumeria 112
Euaspidoceras 43
Euhoplites armatus 41
evolute shells 37, 38, *124*
evolution 21
    biological compromises 54
exogastric coiling 25, (g)150
extinctions 135-41
eyes and eyesight 35
    cuttlefish 90
    nautiluses 28, **66**

falcate ribbing 39
fasciculate ribbing 39
faunal provinces 133
feeding
    ammonites 92-3
    cuttlefish 90
    nautiluses 89-90
    octopuses 90-1
    squids 91-2
foraminifera 92, 93
Frasnian-Fammenian boundary event 135
friction force 84-5
funnel 35
    see also hyponome

geological timescale *135*
gerontic specimens 105
giant squids 21, 48
gills 35, 49-50, 82
Goniatites crenistria 114
Goniatitina (goniatites) 108, 113-15
Gonioclymenia laevigata 112
growth
    influences 104
    phases *100*
    growth lines
        nautiluses 30

growth rates
    nautiluses 103
    tube worms 103-4

haemocyanin 51, (g)151
Hamites 81, *128*
    H. hybridus 43, 44
Haploceras 45
hatchlings
    ammonites *100*
    nautiluses 101
    see also ammonitellas
Hayden, F.V. 45
head 49
'hearts' 51
hectocotylus 95
Hectocotylus 95
helical shells 93, 117
heteromorphs 37, (g)151
    feeding 93
    see also Ancyloceratina
Hildoceras 10, 45
homomorph ammonites 125, 151
hood **66**
    see also operculum
Hoplites 133
Hoplitidae 133
hyponome 16, (g)151
    flexibility 111, 113
    nautiluses 30, **66**
    see also funnel
hyponomic sinus 16, 83
    nautiluses 27

Ichthyosaurus grendelius **67**
internal shells 20, 140
involute shells 36, 37, *121*, *133*
iron pyrites 33, **68**, 143
isotopic shell analysis 104

jaws 47-8
jet propulsion 16, 81-3
Jurassic period ammonites 33, 39, *122*, *124*, *128*
    Lower Jurassic 9, 38, 40, **68-9**, **71**, *121*, *124*, *130*
    Middle Jurassic 29, 77, 96, 97, *129*
    Upper Jurassic 10, 38, 46, 47, **72-3**, **78**, *102*

INDEX

keeled shells 39, *40*, *131*, *133*
Kosmoceras 94
  K. duncani 132
  K. phaeinum 96, 97
K/T boundary extinction event *see* Cretaceous-Tertiary boundary extinction event

lappets 39, **72**, 94, *98*, (g)151
Late Cretaceous period extinctions 137-9
Lechites 95
Lehman, U. 5, 47, 48
lid *see* operculum
lifespan 103-4
Liparoceras naptonensis **71**
lirae 115
living chamber *see* body chamber
lobes 43, *44*, (g)151
  adventitious 120
  diminishing *121*
  divided *118*
  lateral *129*
  pointed *114*
  umbilical *116*, *123*
Loligo forbesi 48
looped ribbing 39
Lytoceras 41, 45, *123*
  L. aaleniarum 42
  L. fimbriatum 124
Lytocerataceae 123
Lytoceratina 109, 122-5

macroconches 23, 78, (g)151-2
  see also sexual dimorphism
mandibles 46, (g)152
  see also beak
manoeuvrability *see* steering and manoeuvrability
Mantell, G. 9
mantle 35, (g)152
  shaping 51
mantle cavity 35
  nautiluses 30
marble, ammonite **80**
mass, centre of 18, 63, 64
maturity *see* age
Meek, F.B. 45
mesoammonoids ('middle ammonites') 108

Mesozoic period extinctions 135-7
meteoric impacts
  and mass extinctions 137-8
microconches 23, **72**, (g)152
  see also sexual dimorphism
'middle ammonites' *see* mesoammonoids
Miocene period ammonites **71**
'modern ammonites' *see* neoammonoids
molluscs 11, **67**
morphology *see* shells
Mortoniceras 95
  M. inflatum 129
mother-of-pearl 101
  see also aragonite
muscle scars 30, *35*, 105
mythological associations 9

nacreous 101
Nautiloidea 12, 15-18, (g)152
Nautilus 23-5
  N. pompilius 16, 25, **65, 66, 79**
  N. umbilicalis 26
nautiluses
  anatomy 25-31
  feeding 89-90
  growth rates 103
  hatchlings 101
  reproduction 98, 99
  size 17
  water pressure resistance 61
neoammonoids ('modern ammonites') 109
Neochetoceras 48
  N. steraspis 47
nepionic constriction 101
neutral buoyancy 14, 55, 58-9, (g)152
Nipponites mirabilis 126

octopuses 13, 140-1
  feeding 90-1
  reproduction 96-7
  see also argonauts
operculum (lid) 45
  nautiluses 27
  see also hood
optic notch
  nautiluses 27
Ordovician period nautiloids 15
orientation 62-4, 81

ornamentation 39-41, 94, 102
orthocone shells 16, 37
ostracods 92, 93
Owen, R. 24
oxycone shells 36, 37, 38, **68-9**, 131
Oxynonticeras 121, 131
   O. lymense 133
   O. oxynotum **68-9**

palaeoammonoids ('ancient ammonites') 108
Palaeozoic period
   extinctions 135-7
   Upper Palaeozoic ammonites 110
paper 'nautiluses' *see* argonauts
Paraceratites elegans 118
Parkinsonia dorsetensis 77, 129
pathology 104-5
patterning *see* coloration and patterning *and* suture lines
Peltoceras dorsetensis 102
pens 140
Perisphinctes 45
Permian period extinctions 119, 135
phragmacone 35, 53, 57, 64, **71**, **76**, (g)152
   nautiluses 28
Phylloceras 45
   P. heterophyllum 121
Phylloceratacae 120
Phylloceratitina 109, 120-2
Physodoceras 47
Pictetia 123, 125
pimples 100
Placenticeras 105, 121
   P. meeki 131
plankton (g)152
Plectronoceras 15, 17
Pleuroceras 47
point force distribution 62
predation
   damage by predators 105
   defence against 62, **71**, 86-7
preservation, modes of 34
pressure, water 59-62
primary constriction 100
prochoanitic septal neck 36
Prolecanites compressus 116
Prolecanitina 108, 115-17

Promicroceras 53
propulsion *see* jet propulsion
prosiphon (g)152-3
protoconches 36, 100, (g)153
pseudolobes 127
Psiloceras 45
   P. planorbis 130

quadrilobate suture lines 120, 127
quinquelobate suture lines 123

radula (g)153
   nautiluses 28
Raup, D.M. 136
religious associations 10
reproduction
   ammonites 99-101
   coleoids 95, 99-100
   nautiluses 98, 99
   octopuses 96-7
   squids 96
retractor muscles 35
   nautiluses 27, 30
retrochoanitic septal neck 36, 109
rhyncholites 28
ribbing 39, 131
   types 39, 102
   with spine bases 127
   see also lirae
rostra 39, 83, 95
rudists 139

saddles 43, (g)153
   diminishing 121
   rounded 114, 118
Scaphitaceae 125, 127
Scaphites 45, 46
   S. nodosus **74**
scarring 30, 89, 105
Schindewolf, O.H. 47, 103
Schlotheimia 103
sea urchins 67
Sepia officianalis 50
Sepkoski, J.J. 136
septa 35, 43, 62, **65**, **68**, **79**, (g)153
   see also suture lines
septal neck 35, 37
   prochoanitic 36

INDEX   159

retrochoanitic 36, 109
serpenticone shells 37, *38*
sexual dimorphism 93-9
   see also macroconches and microconches
shells
   form and function 53-64, 81-7
   growth phases *100*
   morphology 34-41, *100-1*
   oxygen isotopes *104*
   scarring 30, *89*, *105*
   types *37*
   see also internal shells
simple ribbing 39, *102*
siphon 83, (g)153
siphuncle 13-14, 35, 36, 58, **65**, **79**, 111-12, (g)153
size
   nautiluses 17
snakestones 9
soft body anatomy
   ammonites 48-51
   nautiluses 28-30
Solicylmenia 112
spadix 98
spermatophore 95, 98
spine bases *41*
   see also tubercles
spines 39-40, **70**, 86-7, *131*
spinous ribbing *102*
spiral lines **71**
Spirula 55, 57, 61
squids 13, 21, 48, 85
   feeding 91-2
   reproduction 95, *96*
stability 18
steering and manoeuvrability 16, 18
Stephanoceras 45
straight shells
   ammonites **76**
   nautiloids *17*, *19*
streamlining 84-6
suction force 84, 85
suture 35
suture lines 41-4, **77**, *110*, *118*, *129*, *132*, (g)153
   ceratitic *117*, *119*
   complex *116*, *123*, *128*
   goniatitic *114*
   Lytoceratid *124*

Phylloceratid *121*
quadrilobate 120, *127*
quinquelobate *123*
swallowing
   nautiluses 28
swimming
   direction 16-17
   drag 84
   see also jet propulsion

taxonomy and classification *12*, 107-33
teeth 23
teleology 54
Teloceras 9
tentacles 49
Tetragonitaceae *123*
tiger stripe camouflage 24, 50, **71**, 87
   see also coloration and patterning
Titanites titan (T. anguiformis) **78**
Trachyphyllites 123
Tracyceras 21
Trauth 46
Triassic-Jurassic boundary extinctions 136
Triassic period ammonites
   Lower Triassic *118*, *119*
   Middle Triassic *118*
   Upper Triassic *120*
true ammonites see Ammonitina
tube worms
   growth rates 103-4
tubercles 40
   see also spine bases
Turrilitaceae *125*, 126-7
Turrilites 93

umbilicus (g)153
uncoiling 125

vampyromorphs 21
varices 86
ventilation cycle
   nautiluses 30, 83
ventral surface (g)153
Virgatosphinctes sp. *10*
virgatotone ribbing *39*

water displacement 56, 57
water pressure 59-62